Building Materials, Health and Indoor Air Quality

The impact of building materials and construction methods on the health and well-being of occupants is often underestimated. This book is an essential guide to understanding and avoiding hazardous materials and poor air quality in buildings. The author covers a range of issues beginning with an explanation of how buildings work and how this influences the health of occupants and users. The text covers:

- ventilation, air conditioning and indoor air quality
- damp and mould
- asthma and respiratory problems
- cancer and endocrine disorders
- radiation and radon
- hazardous building materials used in construction
- indoor air quality and emissions
- ecological alternatives and approaches and remedies for 'sick' buildings.

The book also guides the reader through the confusing world of regulations, EU and international guidelines and certifications, and provides a critical analysis of different theories of healthy buildings and philosophies.

Written in a clear and accessible style, this book provides indispensable advice and information to anyone wishing to better understand healthy buildings and materials. It is essential reading for architects, surveyors, public health professionals, facilities managers and environmentalists.

Tom Woolley is an architect and environmental consultant in County Down, Northern Ireland. He has been a practising architect and academic, teaching at UK Institutions including the Architectural Association, Strathclyde University, Hull School of Architecture, Queens University Belfast, University of Central Lancashire and the Centre for Alternative Technology. He co-authored the *Green Building Handbook* and is author of *Natural Building*, *Hemp Lime Construction* and *Low Impact Building*.

'As industry and regulation increasingly focus on energy and carbon saving to fulfil our environmental responsibility, this book is a timely reminder that the quality of the buildings we occupy has to be given the same importance as energy performance. In this book Professor Woolley revisits his previous excellent research in toxicity and materials by delivering a book that will be vital in addressing the issues of health and our built environment for practitioners and students alike.'

Alison Pooley, Anglia Ruskin University, UK

'For quite some time issues of the outdoor environment and climate change have overshadowed the issue of indoor air quality. Woolley presents a range of evidence showing that the indoor air quality problem remains highly pertinent, and largely ignored not only by public authorities and the construction industry, but also unfortunately by several health – and environmental organizations.

Modern solutions for energy efficient buildings, such as the passive house, potentially amplify this problem through increased focus on impermeability, new insulation materials and balanced ventilation system with heat recovery.

But Woolley shows us the way forward. He documents with great experience and thoroughness that there is no contradiction between high energy efficiency and good indoor air quality. Proposed solutions include the use of low-emitting materials, hygroscopic surfaces and natural ventilation.

This knowledge is deeply valuable to anyone interested in indoor air quality.'

Bjørn Berge, Gaia Architects

'A powerful plea to make better choices about the products we use and the way we build. This book is a wake-up call for anyone who believes it possible to deliver healthy buildings without challenging the dependence on complex industrial chemicals.'

Gary Newman, Alliance for Sustainable Building Products

Building Materials, Health and Indoor Air Quality

No breathing space?

Tom Woolley

LONDON AND NEW YORK

First published 2017
by Routledge
2 Park Square, Milton Park, Abingdon, Oxon OX14 4RN

and by Routledge
711 Third Avenue, New York, NY 10017

Routledge is an imprint of the Taylor & Francis Group, an informa business

British Library Cataloguing-in-Publication Data
A catalogue record for this book is available from the British Library

Library of Congress Cataloging in Publication Data
Names: Woolley, Tom, author.
Title: Building materials, health and indoor air quality : no breathing space? / Tom Woolley.
Description: Abingdon, Oxon ; New York, NY : Routledge is an imprint of the Taylor & Francis Group, an Informa Business, [2017] | Includes bibliographical references and index.
Identifiers: LCCN 2016016798| ISBN 978-1-138-93447-4 (hardback : alk. paper) | ISBN 978-1-138-93449-8 (pbk. : alk. paper) | ISBN 978-1-315-67796-5 (ebook : alk. paper)
Subjects: LCSH: Buildings—Health aspects. | Sick building syndrome. | Materials—Environmental aspects. | Green products. | Building materials—Environmental aspects.
Classification: LCC RA566.6 .W66 2017 | DDC 613/.5—dc23LC record available at https://lccn.loc.gov/2016016798

ISBN: 978-1-138-93447-4 (hbk)
ISBN: 978-1-138-93449-8 (pbk)
ISBN: 978-1-315-67796-5 (ebk)

Typeset in Times New Roman
by Fish Books Ltd.

In memory of Zane Gbangbola and in support of the Truth about Zane campaign

For more details of the Truth about Zane campaign see pages 63 and 64 in Chapter 5.
See also www.truthaboutzane.com/

Contents

Illustrations

Figures

Tables

Acronyms and abbreviations

AgBB	Committee for Health-related Evaluation of Building Products Germany
AIVC	Air Infiltration and Ventilation Centre
ALARA	as low as reasonably achievable
ARB	Architects Registration Board
ASBP	Alliance for Sustainable Building Products
ASHRAE	American Society of Heating, Refrigerating, and Air-Conditioning Engineers.
ASTM	American Society for Testing and Materials
BB	building biology
BLF	British Lung Foundation
BPA	bisphenol A
BRE	Building Research Establishment
BREEAM	Building Research Establishment Environmental Assessment Methodology
BRI	building-related illness
BSI	British Standards Institute
BUFCA	British Urethane Foam Contractors Association
C2C	cradle to cradle
CAR	Cambridge Architecture Research Group
CAT	Centre for Alternative Technology
CCA	copper chrome arsenic
CE	Conformité Européene
CEN	European Committee for Standardization
CENELEC	European Committee for Electrotechnical Standardisation
ChemSec	International Chemical Secretariat
CHP	combined heat and power
CI	Contamination Index
CIAT	Chartered Institute of Architectural Technologists
CIBSE	Chartered Institute of Building Services Engineers
CIVALLI	Cavity Wall Insulation Victims Alliance
CO	carbon monoxide
CO_2	carbon dioxide
COMEAP	Committee on the Medical Effects of Air Pollution
COPD	chronic obstructive pulmonary disease
CPR	Construction Product Regulations
CRA	condensation risk analysis
CWI	cavity wall insulation

DECC	Department of Energy and Climate Change (UK)
DECO	European Paints Directive
DEFRA	Department of Food and Rural Affairs (UK)
DEHP	di(2-ethylhexyl)phthalate
DIN	German Institute for Standardization
DIY	do-it-yourself
DNEL	Derived No Effect Level
Dy	moisture gradient
EBUK	Earth Building UK
EC	European Commission
ECHA	European Chemicals Agency
EINICS	European Inventory of Existing Commercial Chemical Substances
ELNCS	European List of New Chemical Substances
EMF	electromagnetic field
EPA	Environmental Protection Agency (USA)
EPD	Environmental Product Declaration
EPS	expanded polystyrene
EPSRC	Engineering and Physical Sciences Research Council
EST	Energy Saving Trust
ETIC	external thermal insulation composite
EU	European Union
EUCPR	European Union Construction Product Regulations
EURIMA	European Insulation Manufacturers Association
FACTS	Failure and Accidents Technical Information System
GBH	*Green Building Handbook*
GGBS	granulated blast furnace slag
HBCD	hexabromocyclododecane
HBN	Healthy Building Network
HEAL	Health and Environment Alliance
HEMAC	Health Effects of Modern Airtight Construction
HFC	hydrofluorocarbon
HPA	Health Protection Agency (UK)
HQM	Home Quality Mark
HSE	Health and Safety Executive
HVOTL	high voltage transmission line
IAQ	indoor air quality
IAQA	UK Indoor Air Quality Association
IARC	International Agency for Research on Cancer
IBN	Institute of Building Biology and Sustainability
IDGO	Inclusive Design for Getting Outdoors
IEV	Institute of Environmental Health
III	International Isocyanates Institute
ISIAQ	International Society of Indoor Air Quality and Climate
ISO	International Organization for Standardization
LCA	life cycle analysis
LCI	lowest concentrations of interest
LEED	Leadership in Energy and Environmental Design
MC	moisture content

MDF	medium-density fibreboard
MEARU	Mackintosh Environmental Architecture Research Unit
MIMA	Mineral Wool Insulation Manufacturers Association
MMC	modern methods of construction
MRC	Medical Research Council
MVHR	mechanical ventilation and heat recovery
NGO	non-government organisation
NHBC	National House Building Council
NHL	natural hydraulic lime
NHS	National Health Service
NICE	National Institute for Health and Care Excellence
NIHE	Northern Ireland Housing Executive
NOW	Network of Wellbeing
NTP	National Toxicology Program (USA)
OSB	oriented strandboard
OSHA	Occupational Safety and Health Administration
PAH	polycyclic aromatic hydrocarbon
PAN	Pesticides Action Network
PBDE	polybrominated diphenyl ether
PCB	polychlorinated biphenyls
PCP	pentachlorophenol
PD	Paints Directive
PFA	professional flooring adhesive
PGE	propylene glycol and glycol ether
PH	Passiv Haus
PHPP	Passiv Haus Planning Package
PHF	phenolic foam
PIR	polyisocyanurate
POE	post-occupancy evaluation
POM	particulate matter
PSR	Physicians for Social Responsibility
PU	polyurethane foam
PUR	polyurethane rigid foam
PVC	polyvinyl chloride
RCP	Royal College of Physicians
REACH	Registration, Evaluation, Authorisation and Restriction of Chemicals
RH	relative humidity
RIBA	Royal Institute of British Architects
RICS	Royal Institute of Chartered Surveyors
SAP	Standard Assessment Procedure
SBR	styrene butadiene rubber
SBS	sick building syndrome
SED	Solvent Emission Directive
SIN	Substitute It Now
SIP	structural insulated panel
SPF	spray polyurethane foam
STBA	Sustainable Traditional Building Alliance
SVHC	Substance of Very High Concern

SVOC	semi-volatile organic compound
SWI	solid wall insulation
TCPP	tris 1-chloro-2 phosphate
TDI	toluene diisocyanate
TEAM	Total Exposure Assessment Methodology
TPP	triphenyl phosphate
TPU	thermoplastic polyurethane foam
TSB	Technology Strategy Board (now Innovate UK)
TVOC	total volatile organic compound
UFFI	urea formaldehyde foam insulation
UKBCSD	UK Business Council for Sustainable Development
UKGBC	UK Green Building Council
UKIEG	UK Indoor Environment Group
uPVC	unplasticized polyvinyl chloride
VOC	volatile organic compound
VPU	vapour permeable roof underlay
VTT	Technical Research Centre of Finland
VVOC	very volatile organic compounds
WEN	Womens Environmental Network
WFT	without further testing
WGBC	World Green Building Council
WHO	World Health Organization
WISE	Wales Institute for Sustainable Education
WT	without testing
WWF	World Wide Fund for Nature
XPS	extruded polystyrene

Acknowledgements

Special thanks to:

Lynne Sullivan
Lisa Ponzoni
Rachel Bevan
Leonie Erbsloh

Thanks also due to:

Hannah Bevan-Woolley
Oliver Bevan-Woolley
Peter Groegl
Nigel Kermode
Kavita Kumari
Adrian Leaman
Aisling Macdonald
Cathy Mager
Teresa McGrath
Grainne McGill
Annie Pollock
Ricky Pollock
Pauline Saunders
Mark Singleton
Dr Mohammed Sonebi
Alex Sparrow
Thomas Thorogood
Sophie Woolley
Franz Wortmann
Gabriele Wortmann

Preface

This book launches a broadside attack on those who fail to take account of the true wide-ranging impacts of building design on peoples lives, and will act as a wake-up call to all those engaged with manufacturing, certifying, designing, specifying and installing building products and systems. The author's dedication to the cause of advancement of knowledge of building materials and products is globally recognised, and this latest work finds his appetite for deeper understanding and awareness undiminished. At times controversial, Tom challenges those who make claims for health and wellbeing in their built projects, and asks them to examine the broader context surrounding 'business as usual', including the prevailing commercial and marketing forces, and evaluate from first principles the consequences of specification choices on the quality of peoples lives.

Against the background of leading clients' calls for productivity and health in project outcomes, designers are faced with a shocking lack of agreed standards and legislation in this field and are to a large extent 'flying blind'. As we demand better building performance, our knowledge and understanding of building physics and the chemistry of products and materials needs to increase to an appropriate level – at the very least sufficient to define what 'good' should deliver in terms of the indoor environment in which people spend 90 per cent of their time. Tom asks questions to which we must seek answers, or face consequences which the balance of probability suggests are generally negative for building user health, and potentially disastrous for consumer confidence in the construction sector. The author draws together a powerful argument for urgent action, bringing scientific and medical issues to bear on building design and understanding of real performance in use, and he offers us a range of useful references and pointers to guide us along the path to delivering better indoor air quality and positive outcomes for building owners and occupiers. There are messages here for policy-makers and environmental lawyers as well as building professionals and users – these are timely messages and we must heed the call.

Lynne Sullivan OBE, RIBA

1 Introduction

No breathing space[1]

Public interest in air quality has risen in the past year while this book was being written. Constant press and TV reports about external air pollution have even mentioned indoor air problems and tabloid newspapers have issued dire warnings about scented candles, diesel cars and the return of London smog. However, much of this reporting may have created greater confusion in the minds of professionals and public, rather than raising awareness of pollution risks. This book attempts to unravel the issues in a way that allows the reader to come to their own conclusions about the issues. It builds on the approach we adopted in the 1990s when we produced the *Green Building Handbook* (GBH)[2] in which we attempted to analyse and explain what information was available about the environmental and health risks of building materials and construction methods. This approach was popular with architects and others because it didn't preach or lay down dogmatic answers, though it was not popular with a number of trade organisations such as the British Plastics Federation and similar bodies.

Since then the construction industry in some European countries, such as Germany, Austria, Switzerland and to a lesser extent France and Italy, has moved forward rapidly developing a wide range of sustainable and environmentally friendly certified materials based in part on health impact. The UK and Ireland, however, have been stuck in a suffocating head-in-the-sand mentality that has ignored both the environmental impact and health impact of common forms of construction and has been rigidly fixated on energy efficiency. The buildings that we inhabit, as a result, are contaminated with hazardous and toxic materials that we breathe in every day of our lives and there seems no escape from this. Architects, builders, their clients, government bodies and the medical profession refuse to deal with the environmental causes of ill health and indoor air quality is ignored. Hence the subtitle of this book *No Breathing Space?*, as we seem stuck in a place that we cannot escape from.

Evidence about the health impacts of hazardous construction materials is extensive and compelling, as should become apparent in this book, but we are also trapped as medical epidemiology has largely ignored the issue and evidence of causative effects is patchy at best. We have to rely on circumstantial evidence and make a choice whether to safeguard our health or ignore the risks. This book advocates the precautionary principle as the way to make choices about building materials and products. It warns of the dangers of using synthetic and chemical based materials, when natural and healthy alternatives are available. Evidence of the health risks from hazardous products is set out but none of this is entirely conclusive as much medical epidemiological research into the health effects of toxic and hazardous materials and poor indoor air quality still remains to be done. Overall the evidence about the dangers of hazardous materials is circumstantial and direct association between specific materials and illness cannot always be made. This is because we live in a world where we are breathing in a cocktail of dangerous substances from a wide range of sources on a daily basis. Some of us are lucky enough to avoid any serious illness whereas others succumb to respiratory diseases, asthma or cancer. Pregnant mothers, babies in the womb and young children are particularly vulnerable and subject to endocrine-disrupting substances that are having an impact on reproduction and long-term health.

There are plenty of medical experts and others who like to deny the importance of environmental causes of ill health. They are stuck in a rigid and inflexible approach to healthcare that is based purely on hereditary, genetic and normal clinical explanations for illness. Without any real evidence, they are keen to ignore facts that are staring them in the face about environmental causes for illness. There are also many in the construction industry who are completely wedded to the idea that modern science and technology and the use of chemical materials is the only way to go and they are remarkably prejudiced against the use of safer alternatives. Even lawyers active around environmental issues have failed to engage with indoor air problems. For example, 'ClientEarth has sent a final legal warning to the UK government, which gives the Environment Secretary, Liz Truss, 10 days to act on air pollution or face action in the High Court'.[3,4]

Despite doing excellent work on a wide range of environmental issues, ClientEarth, like many other environmental NGOs, pays little attention to the problems caused by bad indoor air quality: 'While we do share news concerning research into indoor air pollution on our social media channels, we don't do much work on this area. Most of our work focuses on reducing concentrations in outdoor air pollution, such as policies that will reduce traffic emissions'.[5] Sadly the ClientEarth healthy air campaign fails to even mention indoor air quality.[6] This may seem to simply be a case of concentrating resources on one problem, however, as will be shown in Chapter 12, the UK government is using the issue of external air quality as a way of diverting attention from the more serious problem of indoor air.

For many, indoor air quality is an unknown subject but many people have heard of 'sick building syndrome' (SBS).

Whatever happened to sick building syndrome?

It would be reasonable to assume that ensuring good indoor air quality would be a key factor when designing and constructing buildings. The term sick building syndrome became well known in the 1980s and still lingers today, though largely associated with poorly air-conditioned offices. However the health impacts of buildings are still poorly understood and rarely taken into account as a key priority when buildings are designed or renovated. Architects, when specifying building materials are mainly concerned with cost, availability, robustness and aesthetics, not their impacts on health. Yet as we spend up to 90 per cent of our time indoors, the air we breathe is affected by environmental conditions and hazardous emissions from building materials. Indoor air quality (IAQ) is in fact a key issue for our lives.

The aim of this book is to provide, as comprehensively as possible, an introduction as to why many buildings are not healthy, and the contribution of building materials to this and how it is possible to avoid these problems. Hopefully this book will help to throw some light on the real evidence and indicate how important and urgent it is that we reduce the dangers of indoor air pollution.

'The term "sick building syndrome" (SBS) is used to describe situations in which building occupants experience acute health and comfort effects that appear to be linked to time spent in a building, but no specific illness or cause can be identified'.[7] SBS is not a useful concept as it both creates the idea that a building can be sick and also implies that the problem is more with the building occupant (a syndrome) than the building. The UK National Health Service (NHS) explains SBS by listing real symptoms:[8]

- headaches and dizziness
- nausea (feeling sick)
- aches and pains
- fatigue (extreme tiredness)
- poor concentration
- shortness of breath or chest tightness
- eye and throat irritation
- irritated, blocked or runny nose
- skin irritation (skin rashes, dry itchy skin)

As will be seen in this book, it is possible to link illnesses with specific causes in buildings, whether they be damp and mould or toxic pollutants. People who develop illnesses as a result of poor indoor air quality are not always suffering from psychosomatic problems, as was often suggested in the literature about SBS, but are directly affected by pollutants. Recent medical literature, discussed below, also attributes accelerated deaths to pollution.

The symptoms listed above are genuine health problems and many may result from poor indoor air quality and toxic emissions from building materials. It must be acknowledged that there are people who become hypersensitive to certain substances in buildings and this may itself become a health problem in its own right. Some people become extremely anxious about chemical pollution and this may be more severe than the actual medical effects and can be seen as a syndrome, but while serious for the victims, this only affects a tiny number of people. It is important that where readers of this book have anxiety or fears about the issues in this book that they take action to make practical and physical changes to their indoor environment rather than getting ill worrying about it!!

Health problems resulting from or affected by indoor air pollution are considerable and include:

- chronic obstructive pulmonary disease (COPD)
- asthma
- carbon monoxide (CO) poisoning
- respiratory illnesses of various kinds
- irritation
- sensitivity to chemicals and pollutants
- cancer
- hormone disruption
- lethargy from high levels of carbon dioxide (CO_2)

The medical response to indoor pollution

A recent report from the Royal College of Physicians (RCP) and the Royal College of Paediatrics and Child Health, *Every Breath We Take: The lifelong impact of air pollution*, marks a significant change in the attitude of the medical profession to indoor air quality and an acknowledgement of the dangers of chemical pollutants. For many years, environmental causes of illness have been ignored, as genetic, viral and other medical causes have been favoured. More recently, concern has tended to focus on external air pollution but the RCP report at long last recognises the seriousness of indoor air pollution:

> the concentration indoors is as much as ten times higher than that outdoors because of the presence of internal sources. This emphasises the importance of indoor air quality – not only do we spend considerably more time indoors than out, but the range and concentration of pollutants inside buildings are often much greater than those found outdoors.[9]

The RCP states that indoor air pollution may have caused or contributed to 99,000 deaths per annum in Europe but says that the practicalities of setting guidelines and establishing control policies for indoor air pollutants are difficult: 'because of the complexity of pollution sources and the multitude of parties potentially responsible for causing, monitoring and/or regulating indoor air pollution'[10]

Sadly the RCP report still tends to promote the idea that the main problem is with external air pollution from traffic but it does at least go into sources of indoor air pollution including volatile organic compounds (VOCs), formaldehyde, paints, glues, etc., etc. Having drawn attention to these problems they then mistakenly state that while VOCs (the main source of indoor air pollution) are very common 'their health effects are generally minor'. This is poor science, as the RCP does not provide any evidence to back up this statement and even quotes the World Health Organization (WHO) as having classified household air pollution as an 'extremely important risk factor accounting for 4.3 million deaths worldwide in 2012'.[11]

It would be easy to become confused about the health risks of indoor air pollution when reading the RCP report as it plays down the importance of emissions from building materials such as VOCs and puts more emphasis on other emitters such as external air pollution, smoking and radon.

BBC TV and other media reporting of the RCP report was quick to find an 'expert' who was willing to deny the risks of indoor air quality. Professor Anthony Frew was widely interviewed by TV and radio as a critic of the report, suggesting that improvements in air quality would only save a few lives and that deaths from air pollution had only been 'brought forward'. In other words, people affected would have died anyway! Frew believes that the cost of improving air quality would not be worthwhile and this reflects the attitude among much of the healthcare world that it is worth putting up with the risks from modern industry and technology:

> Living has never been risk-free and we make compromises all the time between our short-term comfort and long-term health … Its findings have to be seen in the context that on average we live longer, healthier lives than we did in previous generations, and that much of this is due to the industrial improvements that cause the pollution.[12]

Frew (a consultant at the Royal Sussex County Hospital and a Nuffield Health consultant) working in respiratory medicine, asthma and allergy treatment and immunotherapy represents the mainstream medical view, which constantly downplays the significance of environmental causes, raising doubts about environmental improvements because of the cost, and instead preferring drug treatments. His comment about short-term risks will not be much comfort to those suffering from severe respiratory conditions or cancer.[13,14]

> Allergen avoidance: once an allergic disorder had developed, the chances of further attacks could be minimised by avoiding the allergen or other environmental conditions which aggravated the condition. However, the question of whether the avoidance of factors such as house dust mite resulted in substantial clinical improvement, whether these approaches were cost-effective, and who should pay for such measures, was controversial.[15]

When a medical expert says something is controversial it is a coded way of saying that they don't agree with it. There is the implication that environmental solutions would be too costly, however the drug treatments Frew advocates are also very expensive to develop and may have unwanted side effects. There are strong vested interests in promoting the multi-million-pound drugs industry rather than searching for simpler and cheaper environmental solutions. The medical experts create win–win situations for themselves by failing to carry out epidemiological studies of environmental causes of illness and then saying that there is insufficient evidence!

The construction industry's attitude to indoor pollution

By and large, the construction industry has a head-in-the-sand attitude to indoor pollution and air quality. When carrying out research for this book, both clients and their architects were interviewed about their approach to healthy buildings and it was found that general ignorance was the norm. When offered the chance to learn more about the subject and commission a low-cost IAQ test of a building, the offers were declined without exception. Of course, IAQ problems can rarely be seen (except in the case of mould growth) and thus 'out of sight out of mind' may have been the general view. Even architects involved in the design of healthcare buildings seemed unaware of the issues.

A lack of awareness of IAQ is reflective of the lack of post-occupancy evaluations of buildings. Rarely does anyone check whether the building achieved what was intended.

> Designers and builders are trained to undertake building work and hand over the keys, not to look into what happens afterwards. Few clients want to pay for anything more. Meanwhile government, which helped to close the feedback loop via its building design, management and research departments, … tends to have outsourced, privatized or abandoned most of these functions.[16]

The lack of post-occupancy evaluation (POE) means that little is known about whether buildings ever achieve the energy and environmental conditions predicted, and nothing is checked about the IAQ. Somehow as a society we have come to accept poor-quality buildings; just something to grumble about along with the weather. However the scientific evidence that does exist confirms that many buildings have serious performance problems, heating bills are higher than expected, predicted insulation performance unachieved, mould and damp are widespread and occupant comfort is problematic. 'Ahhh....the comforts of home. If a home is "efficient," surely it's "comfortable," right? Not necessarily'.[17] The poor performance of modern buildings has been well documented and this is often referred to as the 'performance gap'. Mainstream organisations such as Zero Carbon Hub and Innovate UK[18] have been surprisingly open about the poor performance of the rare examples of POE tested buildings, though they have largely been concerned with energy efficiency, not IAQ.[19]

Some expert studies have drawn attention to concerns about IAQ but research funding bodies have also followed the head-in-the-sand approach and instructed scientists not to stray too far into this territory. Some academics have reported to me that they were told that the Technology Strategy Board (TSB) (now Innovate UK) in POE studies would not fund a full spectrum analysis of IAQ problems, over and above CO_2 levels. IAQ analysis, including measurement of VOCs, is mentioned in the TSB criteria for POE studies, but only as a discretionary item that would not be funded.[20]

Over a hundred POE projects have been financed by TSB/Innovate UK, but as far as can be determined, these have ignored IAQ issues. An early-findings report gives a brief mention of CO_2 in relation to a ventilation system but despite the considerable amount of public money devoted to these projects it is very hard to access the results. A few case studies were found but with very brief superficial reports and none dealing with IAQ or related issues.[21]

During the research for this book it became clear that in the UK, in particular, official bodies in government, public health and the construction industry have decided to ignore IAQ and healthy building issues almost entirely. Indeed they have tried to distract attention by focusing on external air quality and funding research into concepts of 'healthy place-making' and 'wellness and wellbeing', rather than healthy buildings. This is discussed in more detail in Chapter 12. This is an important issue as the construction industry is unlikely to address IAQ issues unless forced to do so by government policy or regulations. The UK government has focused resources and publicity on external rather than indoor air pollution and the UK Department of Food and Rural Affairs (DEFRA) guide to air pollution does not mention IAQ once.[22]

There is no doubt that traffic and chemical pollution outdoors have a bad effect on our health and emissions should be reduced, but many reports on this problem fail to mention indoor air pollution even though, as will be shown later, this is a much more serious problem. 'More people die from air pollution than Malaria and HIV/Aids, new study shows: More than 3 million people die prematurely each year from outdoor pollution and without action deaths will double by 2050'.[23]

Professional ethics fail to take account of healthy buildings

It might be expected that professional ethics would require architects, surveyors and engineers to pay attention to the health of building occupants but this is far from the case. The Royal Institute of British Architects (RIBA) Code of Conduct merely states '(3.2) Members should be *aware* of the environmental impact of their work'.[24] The Architects Registration Board (ARB) code is also very weak on environmental responsibility: '5.1 Whilst your primary responsibility is to your clients, you should *take into account the environmental impact* of your professional activities.' However there are proposals to revise the code and currently a slightly better revision is being considered, but at the time of writing, not yet adopted. However even the new improved wording is fairly weak as it does not refer to environmental or health responsibilities for buildings: '5.1 Where appropriate, you should advise your client how best to conserve and enhance the quality of the environment and its natural resources'.[25]

The Royal Institute of Chartered Surveyors (RICS) makes no reference to environmental or health responsibilities in its code.[26] However the Chartered Institute of Building Services Engineers (CIBSE) does include the following: '6. Have due regard to the safety, health and welfare of themselves, colleagues and the general public. 7. Have due regard to environmental issues in carrying out their professional duties'.[27] The CIBSE also provides a definition of IAQ: '2.1 Definition and importance of IAQ 2.2 Why ventilation is required 2.3 Requirements for good IAQ 2.4 Regulations and standards 2.5 Common pollutants, pollutant sources and related health issues 2.5.1 Pollutant types: Gaseous pollutants, Volatile organic compounds (VOCs), Odours, Particulates, Water Vapour'.[28]

Indoor air quality is more serious than external air pollution

Despite official attempts to ignore IAQ in the UK, there are many authoritative sources that point out that indoor air problems have a much more significant effect on our health than external pollution:

> indoor air is often more polluted than outside. Many factors contribute to poor indoor air quality, including tobacco smoke; cleaning products; pesticides; formaldehyde; cleaning agents; waxes and polishing compounds; fragrances; plasticisers in wallpaper, rugs and fragrances; components of building structures (such as sealants, plastics adhesives, and insulation materials).[29]

> Indoor air pollution is often higher than outdoor air pollution. This has been observed in indoor spaces including public buildings, schools, kindergartens and homes in developed countries, especially when significant sources of indoor and outdoor air pollution exist. The quality of the indoor environment is affected by the quality of ambient air; building materials and ventilation; consumer products, including furnishings and electrical appliances, cleaning and household products; occupants' behaviour, such as smoking; and building maintenance. For example, energy saving measures make buildings and houses airtight by reducing ventilation, thus raising concerns over indoor air quality, as chemical and biological pollutant concentrations can reach high levels.[30]

Most medical organisations have ignored IAQ, with the notable exception of the British Lung Foundation that has published some reasonable, though slightly flawed, guidelines: *Your Home and Your Lungs*: 'You might consider using building materials with low emissions. Look for products and materials that show they are environmentally friendly and low in pollutants and emissions'.[31] Cancer charities, rather surprisingly do not have policies of excluding carcinogenic materials from their own buildings, and when interviewed showed little knowledge or understanding of indoor air problems, as discussed more fully in Chapter 4.

It will be essential to win over the medical professionals and scientists to understanding building-related medical problems, as suggested by Zhang:

> small changes in indoor pollutant sources can have the equivalent health benefit as large changes in outdoor sources for the same pollutant. Without the active participation of building designers who appreciate the importance of indoor air quality, it would be impossible to design and construct healthy buildings. Likewise, without the active participation of health scientists and professionals in resolving the puzzles of building-related medically unexplained symptoms and other illnesses, it would be impossible to develop reliable and effective guidelines to prevent excess health risks associated with poor indoor air quality.[32]

The issues of airtightness, energy efficiency and problems with ventilation are explored more fully in this book, as this may have unwittingly caused an increase in bad IAQ.

This leads us back to the issue of SBS because it was frequently put down to poor ventilation. SBS is a term familiar to many people, but was largely seen as an issue for mechanical and electrical engineers responsible for ventilation and air conditioning. NHS guidance suggests that SBS is poorly understood though there is also some truth in the following:

> Anyone can be affected by SBS, but office workers in modern buildings without opening windows and with mechanical ventilation or air conditioning systems are most at risk … Since the 1970s, researchers have tried to identify the cause of SBS. As yet, no single cause has been identified. Most experts believe that it may be the result of a combination of things.[33]

The US Environmental Protection Agency (EPA) issued a fact sheet on SBS in 1991: 'A 1984 World Health Organization Committee report suggested that up to 30 percent of new and remodeled buildings worldwide may be the subject of excessive complaints related to indoor air quality (IAQ)'.[34]

SBS is a misleading term, as it implies that buildings can make you sick, whereas the problem is much more to do with poor IAQ, and not necessarily the whole building. Quite minor changes in materials may remove the problem. SBS has largely faded from the headlines and professional concerns in recent years. 'What Happened to Sick Building Syndrome? SBS Has Faded from Headlines, But Is It Still a Problem?'.[35] However problems of building occupants feeling sick have not gone away and with increased air tightness and the growing use of hazardous materials, problems may be getting worse. 'Is Your Office Killing You? – Sick Building Syndrome – Is it Still a Problem?'[36]

SBS has been redefined as building-related illness (BRI) as a more useful term:

> BRI is generally an allergic reaction or infection, and specific symptoms include cough, chest tightness, fever, chills, and muscle aches. The symptoms may continue after occupants have left the building, and the cause of symptoms is known,' said Stadtner. 'Humidifier fever, legionnaires' disease, skin rashes, hypersensitivity pneumonitis, and other illnesses related to bacteria, fungus (mold), and viruses are often classified as BRI, not SBS.[37,38]

Choosing less-hazardous building materials

One of the main contributors to poor IAQ (apart from ventilation problems) is the way in which the building is constructed and the materials used to construct it. Thus this book largely focuses on what is known about construction materials and how these might create health problems. Rather than stating that certain materials are good or bad, this judgement is left to the reader to base their choice and specification decisions on the best information that it has been possible to present here.

If in doubt, the best approach to adopt is that of the *precautionary principle*. If there appear to be serious concerns, even if reliable scientific information still leaves some doubt, then surely it is better to avoid the risk and go for an alternative that is substantially less risky. Fortunately today there are plenty of building methods and materials available that are very low risk and largely avoid the use of hazardous chemicals.

Unfortunately the industry has relied for some years on the *Green Guide to Specification*,[39] produced by the Building Research Establishment (BRE), to select materials on the misapprehension that this provides a good environmental guide. Unfortunately the *Green Guide* has largely ignored health, and embodied energy in its life cycle analysis (LCA) methodology. Any product that was alleged to contribute to energy efficiency was given an 'A star' rating, whether made from hazardous materials or not.[40,41] Fortunately the BRE has introduced IAQ standards into the Building Research Establishment Environmental Assessment Methodology (BREEAM) and the new Home Quality mark but there are real contradictions with guidance about materials in the *Green Guide*.

Structure of the book

In Chapter 2 the nature of VOCs is explained, where they come from and how they can be measured in buildings. Reducing VOC emissions is a crucial step towards healthier buildings. As VOC emissions are almost entirely from synthetic chemical compounds, in Chapter 3 the origins of chemicals are explored. The question is asked, why do we need to be so dependent on chemicals and synthetic products? The chapter also gives examples of the negative impact on the environment and people's lives from the manufacture of such materials. Cancer and formaldehyde are discussed together in Chapter 4, as carcinogenic and mutagenic materials are some of the most dangerous and worrying in building materials. Then in Chapter 5 the many other hazardous aspects of buildings are discussed, including electromagnetic radiation and radon and how these are also a problem of building materials.

In Chapter 6 an attempt is made to summarise the most dangerous materials and products and why these should be avoided. Mould, damp and condensation are an intractable problem in buildings and continue to remain a problem despite efforts to make better designed and constructed energy-efficient buildings. Despite efforts to deal with fuel poverty, many people still live in cold damp houses and even where buildings have been insulated the problems do not go away and even get worse. In Chapter 7 the reasons for dampness and its relationship with building materials are explored. Closely related to damp and mould is the issue of ventilation as this is a crucial factor in IAQ. However in Chapter 8 it is explained that simply using more and more mechanical ventilation is not necessarily the best approach. Dependence on mechanical systems such as those advocated in Passiv Haus can lead to serious health concerns.

Having discussed all the things that can go wrong, possible solutions to create healthier buildings through renovation are set out, with suggestions as to suitable materials and products that do not contain hazardous chemicals and can have other beneficial health effects, including ways to improve existing buildings where hazardous materials are already in place. The lack of a sound theory of design for healthy buildings is a theme in Chapter 10, though there are a number of different approaches that may suggest ways forward. In Chapter 11 the different options and materials to build healthier buildings are set out, including emphasising what should be avoided. Finally in Chapter 12 government and European policies and regulations are explained that, together with the lack of a cohesive approach, make it difficult to ensure good IAQ in the near future.

Notes

1 Richard Pollock, Hammond Care Dementia Centre

2 Woolley, T., Kimmins, S., Harrison, P. and Harrison, R. (1997 & 2000) *Green Building Handbook*, Volumes 1 and 2, E&FN Spon

3 www.clientearth.org/news/latest-news/final-legal-warning-gives-government-10-days-to-act-on-air-pollution-3166 [viewed 1.2.16]
4 'Law firm in new legal threat over UK air pollution' www.bbc.co.uk/news/uk-35689427 [viewed 1.3.16]
5 Personal email from Gavin Thomson, ClientEarth, 14 December 2014
6 http://healthyair.org.uk/documents/2014/11/healthy-air-campaign-policy-call-2.pdf/ [viewed 3.3.16]
7 United States Environmental Protection Agency (1991) *Indoor Air Facts Sheet*, Number 4, November, EPA
8 www.nhs.uk/Conditions/Sick-building-syndrome/Pages/Introduction.aspx [viewed 3.3.16]
9 Royal College of Physicians (2016) *Every Breath We Take: The lifelong impact of air pollution*. Report of a working party. RCP.
10 Royal College of Physicians (2016) op cit.
11 World Health Organization (2010) *WHO Guidelines for Indoor Air Quality: Selected pollutants*. WHO. www.euro.who.int/__data/assets/pdf_file/0009/128169/e94535.pdf [viewed 10.3. 2016]
12 http://foreignaffairs.co.nz/2016/02/23/expert-reaction-to-new-report-on-the-lifelong-impact-of-air pollution/#sthash.fm4ubKkk.dpuf [viewed 1.3.16]
13 www.nuffieldhealth.com/consultants/professor-anthony-frew [viewed 25.1.16]
14 www.bsuh.nhs.uk/departments/respiratory-medicine/the-team/ [viewed 25.1.16]
15 www.publications.parliament.uk/pa/ld200607/ldselect/lds [viewed 10.3.16]
16 Bordass B. and Leaman A. (2013) A new professionalism: remedy or fantasy? *Building Research & Information*, 41 (1), 1–7
17 Grisolia A. (2015) Comfort is measured by more than energy efficiency, *Builder* www.builder online.com/building/building-science/indoor-comfort-is-measured-in-more-than-just-energy-efficiency_o [viewed 25.1.16]
18 Innovate UK (TSB), Building performance evaluation, *Connect*. www.connect.innovateuk.org/web/building-performance-evaluation [viewed 25.1.16]
19 Zero Carbon Hub (2014) Summary: The performance gap: End of term report www.zerocarbonhub.org/sites/default/files/resources/reports/EndofTerm_Summary.pdf [viewed 25.1.16]
20 Technology Strategy Board (2012) *Building Performance Evaluation: Domestic Buildings Guidance for Project Execution*, February 7, TSB
21 Innovate UK (2014) Building Performance Evaluation programme Early Findings from Non-Domestic Projects, https://connect.innovateuk.org/web/building-performance-evaluation/case-studies [viewed 1.3.2016]
22 Department of Environment, Food and Rural Affairs (2014) *Guide to UK Air Pollution Information*, June 2014 http://uk-air.defra.gov.uk/air-pollution/ [viewed 25.1.16]
23 www.theguardian.com/environment/2015/sep/16/more-people-die-from-air-pollution-than-malaria-and-hivaids-new-study-shows [viewed 25.1.16]
24 www.architecture.com/RIBA/Professionalsupport/Professionalstandards/CodeOfConduct.aspx [viewed 14.3.16]
25 http://ebulletin.arb.org.uk/february2016/dear-architect-your-code-your-say/ [viewed 14.3.16]
26 www.rics.org/us/regulation1/rules- of-conduct1/ [viewed 14.03.16]
27 CIBSE Council (1995) *Code of Conduct*, www.cibse.org/Society-of-Public-Health-Engineers-SoPHE/Code-of-Conduct [viewed 25.1.16]
28 KS17 Indoor Air Quality and Ventilation, CIBSE www.cibse.org/Knowledge/CIBSE-KS/KS17-Indoor-Air-Quality-Ventilation [viewed 25.1.16]
29 Wargo, J. (2009) *Green Intelligence, Creating Environments that Protect Human Health*, Yale University
30 European Environment Agency (2013) *Environment and Human Health*, EEA Report No 5/2013, Report EUR 25933 EN, Joint EEA-JRC report, EEA
31 www.blf.org.uk/support-for-you/your-home-and-your-lungs/causes/building-materials [viewed 20.3.16]

32 Zhang J. J. and Smith K. R. (2003) Indoor air pollution: a global health concern, *Oxford Journals Medicine & Health British Medical Bulletin*, 68 (1), 209–225
33 NHS, *Sick Building Syndrome* www.nhs.uk/Conditions/Sick-building-syndrome/Pages/Introduction.aspx [viewed 25.1.16]
34 EPA *Indoor Air Facts No. 4 Sick Building Syndrome*. www.epa.gov/indoor-air-quality-iaq/indoor-air-facts-no-4-sick-building-syndrome [viewed 20.3.16]
35 www.highbeam.com/doc/1G1-369177937.html [viewed 20.3.16]
36 Levine L., CIH, LEED info@ecothinkgroup.com, www.ecothinkgroup.com/is-your-office-killing-you/ [viewed 20.3.16]
37 Turpin J. (2014) What happened to sick building syndrome?, *The News* www.achrnews.com/articles/126576-what-happened-to-sick-building-syndrome [viewed 26.1.16]
38 http://healthybuildingscience.com/2013/07/04/sick-building/ [viewed 20.3.16]
39 www.bre.co.uk/greenguide/podpage.jsp?id=2126 [viewed 20.3.16]
40 May N. and Newman G. (2008) *Critique of Green Guide to Specification*, The Good Homes Alliance, 5 December
41 The Alliance for Sustainable Building Products (2011) *The BRE Green Guide to Specification*, submitted to UKGBC 30 December

2 Volatile organic compound emissions

Volatile organic compounds are the principal source of hazardous indoor air emissions. It is easy to refer to VOCs, but what are they, and which are the most problematic and how can they be controlled?

VOCs include formaldehyde and other similar materials, though formaldehyde is dealt with in more detail in a Chapter 4. VOCs are generally seen as irritants whereas aldehydes can be classified as carcinogenic. However VOCs can also be classified as mutagenic and some may convert to aldehydes through interaction with ozone. Mutagens are 'genotoxic' in that they can change the genetic material of an organism and increase the likelihood of mutations.

According to the WHO and the US EPA, the average levels of several organic compounds in indoor air are two to five times higher than in outdoor air. During certain activities, such as paint stripping, and for several hours immediately afterwards, levels may be a *thousand times* higher than outdoor levels.[1]

What is a VOC?

The term VOC refers to carbon based chemicals that are present in a gaseous form at room temperature. CO_2 and CO are *inorganic* gases and are not classified as VOCs. VOC gases can be formed by a wide variety of molecules with different structures and forms. There are also semi-volatile organic compounds (SVOCs), which are particularly important in the context of this book and include plasticisers, flame-retardants and pesticides. VOCs exist not only as gases but as chemicals adsorbed onto indoor surfaces. SVOCs are largely present on surfaces and on particles rather than in a gaseous form.

VOCs are measured in parts per billion (ppb) or parts per million (ppm) and most commonly micrograms per cubic metre ($\mu g/m^3$). A microgram is one millionth of a gram. If the concentration is 1 $\mu g/m^3$, then for every cubic metre volume of air there is 1 microgram of mass (weight) of the VOC.

This might seem to be a tiny amount, and generally VOCs and SVOCs are present in low concentrations, but higher concentrations of VOCs are considered to be a serious health risk and as a result, many countries have established 'safe' VOC levels. If concentrations exceed these safe levels then action should be taken to reduce VOC emissions (though this is not always very easy to do). 300 $\mu g/m^3$ is commonly used as a safe level but some VOCs can be present at a higher rate than 1000 $\mu g/m^3$.

Some people can be hypersensitive to chemicals such as VOCs at much lower levels and once sensitised maybe affected at levels below what are given as safe levels. 'Safe levels' should be regarded as guidance rather than a rigid concept.

The source of some VOCs can be from external air but indoor air concentrations of VOCs far exceed external sources. Individual VOCs can be measured through IAQ testing but very often tests refer to total volatile organic compounds. The use of TVOCs can be a cheaper way of measuring VOC levels, though it is not always very accurate or sufficiently detailed. Some government standards are based on TVOC levels rather than individual VOC emissions. TVOC readings will indicate that there are problems from a range of individual VOCs but will not necessarily help with identifying the source of the problem. VOCs can be grouped into four categories according to Crump (including particulates):

> Organic compounds that originate from a variety of indoor sources and outdoors. Hundreds of organic chemicals occur in indoor air although most are present at very low concentrations. These can be grouped according to their boiling points … four groups of organic compounds: (1) very volatile organic compounds (VVOC); (2) volatile (VOC); (3) semi-volatile (SVOC); and (4) organic compounds associated with particulate matter (POM). Particle-phase organics are dissolved in or adsorbed on particulate matter. They may occur in both the vapor and particle phase depending on their volatility. For example, polyaromatic hydrocarbons (PAHs) consisting of two fused benzene rings (e.g., naphthalene) are found principally in the vapor phase and those consisting of five rings (e.g. benz[a]pyrene) are found predominantly in the particle phase.[2]

Testing for VOCs and other indoor air quality problems

There are a number of businesses that can carry out IAQ tests in the UK. The bigger commercial companies get most of their business from factories where chemicals are being produced and health and safety rules require emission levels be kept under control. They will also test hospital operating theatres and other places where there are specialised needs in terms of air quality. Some universities can also carry out analysis work but this is likely to be quite expensive. In recent years simpler and more affordable home survey test kits and methods have become available and thus, for a few hundred pounds it is possible to check on IAQ in a home, school or office. The following is extracted from a home air test report:

> What's in your Indoor Air Quality Report?
> Your Indoor Air Quality Report has several sections describing different aspects of your home's air quality.
>
> 1 The Total Volatile Organic Compound (TVOC) level: a general indicator of the IAQ in your home. Typically, a lower TVOC means better IAQ in your home.
> 2 The Total Mould Volatile Organic Compound (TMVOC) level: an assessment of the actively growing mould in your home. Levels above 8 μg/m^3 indicate that there is a source of actively growing mould in your home.
> 3 The Contamination Index™ (CI): shows the types of air-contaminating products and materials that are present in your home. Each CI category shows the approximate contribution of that category to the TVOC level, indicates how your home compares to thousands of other homes, and provides some suggestions for where these products and materials might be found. The CI is divided into 3 main sections: Building-Related Sources, Mixed Building and Lifestyle Sources, and Lifestyle Sources. Building-Related Sources are those that are typically part of the structure of the home and may be more difficult to reduce in the short term.

Mixed Building and Lifestyle Sources are those that could belong to either category and investigation on your part may be necessary to determine which source is more likely. Lifestyle Sources are those that the occupants of the home bring into the home and can usually be readily identified and remediated. Levels indicated as Elevated, High, or Severe should be immediately addressed, and those listed as Moderate are areas that can be improved over time. Since there are potentially many sources of VOCs, homes can often be re-contaminated even after sources have been removed because new products are constantly being brought into the home. Home occupants and homebuyers should take note of this fact, and view IAQ as a continuous improvement process.

4 Significant VOCs: listing of the chemical compounds measured with the IAQ Home Survey test that are large contributors to the TVOC level or are indicative of specific types of products or problems. Reduction of these specific chemical compounds will substantially reduce the TVOC level and greatly improve the IAQ of the home.

Total Volatile Organic Compound (TVOC) Summary

Your TVOC Level is (μg/m³): 940 **HAC Air Quality level: Moderate**

Approximately 8,000 Samples
Median TVOC (μg/m³): 1100
(midpoint value where half the points are above this value and half are below)
Mean or Average TVOC (μg/m³): 1900
(sum of all values divided by the number of values)

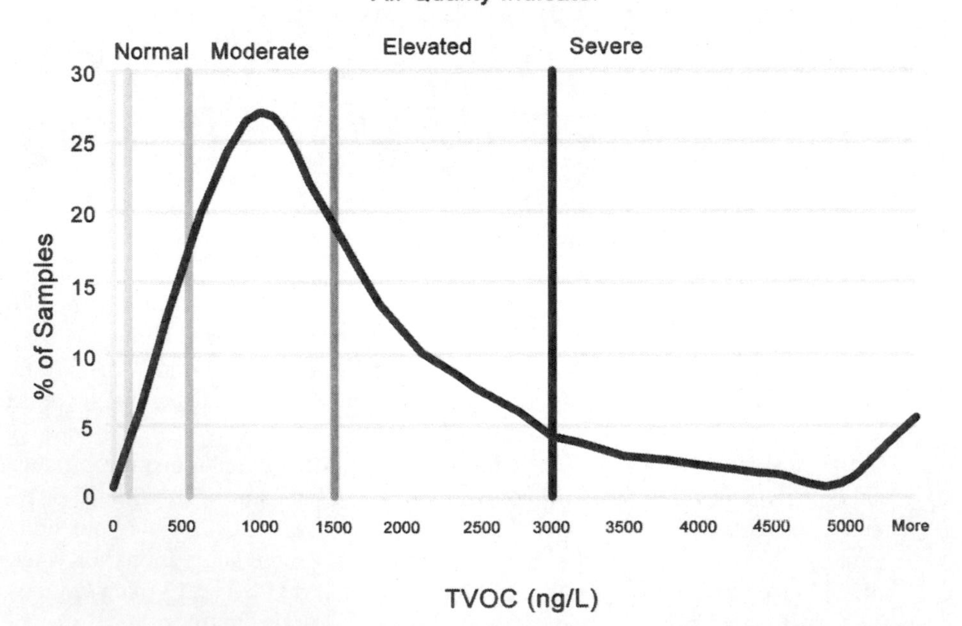

Figure 2.1 Typical total volatile organic compound (TVOC) summary

Source: Reproduced with the permission of Waverton Analytics and Prism Analytical Technologies[3]

5 EPA Hazardous Air Pollutants (HAPs): listing of the chemical compounds measured with the IAQ Home Survey test that are known or suspected to have serious health or environmental effects (also known as air toxics).

There is a wide range of detection methods from highly sensitive and expensive through to quite small and relatively affordable air sampling kits. Air sampling involves a range of devices that can absorb air from a room-sized space. Testing normally involves using a small pump, which is placed in one or more rooms, for a prescribed period of time. This samples the air, which then goes to a laboratory where it is analysed using gas chromatography/mass spectrometry. Laboratory analysis should be carried out to International Organization for Standardization (ISO) standards or other standards such as those set by the German Institute for Standardization (DIN) or the American Society for Testing and Materials (ASTM). There are also accreditation schemes for laboratories. Testing methodology is a vast science in itself and more needs to be done to standardise testing and the way in which results are presented to avoid confusion and for comparisons to be made between buildings.

It is possible to get some information about VOCs and other emissions from a range of new handheld devices and personal monitors that are coming onto the market, though they may not be as sensitive or accurate as (pumped) air sampling. Some use photoionisation, which gives a fast response but will not necessarily give an accurate picture of emissions in a space over a period of time and has been developed instead to give warnings to workers in a higher risk situation. There are also IAQ monitors and data loggers that can be stationed around a building and log emission data from a wide range of toxic chemicals.[4,5,6]

Measuring VOC emissions

There has been a great deal of scientific work concerned with measuring building material emissions rates. One paper dealing with diffusion coefficients points out that there can be a variation of 700 per cent between results using different methods, and so depending on the methodology used there can be an underreporting of emissions.[7]

Adsorption of materials is used in laboratory experiments to measure how VOCs are emitted. Such laboratory research is valuable because it confirms how easily dangerous levels of chemicals are released. Commercial companies that are responsible for materials that produce VOCs and other emissions tend to take the line that emission levels are so low that there is not really a serious health problem. However the scientific literature on what are known as ‘secondary emissions’ confirms that emission rates can be surprisingly high even when a material has been in place in a building for some time. Emission rates do fall, especially if ventilation rates are high, but this may not be the case in increasingly airtight buildings.

> Most inner surfaces may store chemicals and equilibrium with the surrounding air will be approached by desorption or adsorption of airborne chemicals … adsorbed pollutants may diffuse into materials and their reemission is typically slower and also influenced by the speed of diffusion back to the surface.[8]

As any materials will adsorb/absorb VOC emissions, this leads on to the concept of chemical sinks, where hazardous chemicals are stored within the building fabric. This has been suggested as a useful way to deal with the problem but chemicals can be re-emitted. Spiegel and Meadows for instance have pointed out that all surface materials will absorb chemical molecules but then suggest that these are then released again later.[9]

The health effects of VOCs

These are normally referred to as:

- Acute:
  - Irritation of eyes and respiratory tract
  - General: headache, dizziness, loss of coordination, nausea and visual disorders
  - Allergic reactions, including asthma and rhinitis
- Chronic:
  - Damage to liver, kidney, blood system and central nervous system
  - Some may cause cancer in humans

The evidence about health problems caused by VOCs can be very mixed and confusing. Medical epidemiology in this area is limited and there appear to have been insufficient studies carried out where IAQ measurements in buildings have been correlated with VOC emissions and health impacts. Some work has been done in laboratory-controlled experiments however.

> In one series of experiments, humans were exposed to concentrations of a specific mixture of 22 VOCs typically occurring in indoor air … and known to be emitted from building materials. In the experiments, where the subjects exposed were humans who previously had felt SBS-symptoms, a number of subjective reactions and neuro-behavioural impairment occurred at TVOC concentrations of 25 mg/m^3 and odour appeared at 5 mg/m^3, which was the lowest concentration used in these experiments. The effects occurred within minutes after the start of exposure.[10]

> In a controlled chamber study the reactions of 21 healthy persons were compared with a group of 14 persons suffering from the sick building syndrome (SBS subjects) when exposed to 25 mg/m^3 of the same specific mixture of 22 VOCs as above. A tendency to a stronger response was seen among the SBS subjects. Physiological measures indicated exposure-related reduction of lung function among the SBS persons.[11]

Scientific research into the effects of VOCs on health carried out in controlled laboratory situations usually creates an exposure to VOCs for a fairly limited period of time. While such experiments can control variables and provide useful data, they barely begin to reproduce the conditions in a home where occupants are exposed to a wide range of VOCs over a period of months or years.[12]

Much of the general literature appears to suggest an uncertainty about the links between VOCs and health problems: 'While some studies reported positive correlation between symptoms and total VOC concentration others did not. Several studies even showed a *reduced* prevalence of symptoms with increasing VOC-concentrations'.[13]

While Sundell, a leading authority in this field, seemed uncertain about the link between VOC emissions and health problems in 1993, he stated in 2004:

> that in the developed world IAQ is a main cause of allergies, other hypersensitivity reactions, airway infections, and cancers. Allergies, airway infections and sick building syndrome are associated with, e.g., "dampness", a low ventilation rate, and plasticizers. In the future more emphasis must be given to IAQ and health issues.[14]

Earlier medical literature in the 1990s tended to play down the significance of VOCs and IAQ as a health problem but more recently this has changed, particularly in view of the increasing airtightness of buildings. A 2003 paper in the *British Medical Bulletin* warns of the dangers of exposure to chemicals and makes a clear link between increasingly airtight buildings, construction chemicals and health problems:

> Everyday exposure to multiple chemicals, most of which are present indoors, may contribute to increasing prevalence of asthma, autism, childhood cancer, medically unexplained symptoms, and perhaps other illnesses. Since the 1970s … Improved energy conservation was mainly achieved through reducing exchanges between outdoor fresh air and indoor air. Meanwhile, synthetic materials and chemical products have been extensively used in these airtight buildings. The combination of low ventilation rate and the presence of numerous sources of synthetic chemicals has resulted in elevated concentrations of volatile organic compounds (VOCs) (e.g. benzene, toluene, formaldehyde), semi volatile organic compounds (SVOCs) (e.g. phthalate plasticizers and pesticides) and human bioeffluents. This has been suggested as a major contributing factor to occupant complaints of illness symptoms, or so-called 'sick building syndrome', in the last three decades. Although the aetiology is still not clear, many cases of respiratory diseases, allergies and asthma, medically unexplained symptoms including sick building syndrome, and cancer, are believed to be attributable to poor indoor air quality in both developing and developed countries.[15]

It is relatively easy for those wishing to detract from concern about healthy building issues to claim that there is insufficient medical confirmation of health and VOC links, but as Zhang and Smith confirm, the risk is too high to ignore these issues. Health problems related to VOC emissions are also of concern to the insurance industry, anticipating future claims and litigation as 'reducing volatile organic chemical emissions from products is a critical aspect of protecting human health'.[16]

> In recent years indoor air quality (IAQ) problems leading to the so-called 'sick building syndrome' have fed considerable litigation. Personal injury claims, which utilize remedies under insurance disability and workers' compensation laws, or the Occupational Safety and Hazard Act, as well as property damage, constructive eviction, and design and construction defect claims may relate to IAQ problems … Pollution exclusion clauses exclude from coverage emissions that may be the cause of IAQ problems. Since the mid-1980s, insurers have attempted to avoid the costs of environmental pollution by the use of pollution exclusion clauses, and have placed IAQ claims under the same exclusionary blanket as 'traditional' environmental pollution sources in order to avoid payment on these claims. Court decisions are mixed as to whether insurance companies can use these clauses to refuse coverage of IAQ-related claims.[17]

Some of the literature contains the assumption that VOC emissions have significantly reduced in developed countries due to restrictions on the use of solvents and VOCs. This assumption is possibly over optimistic but a secondary threat has been diagnosed from the interaction of certain chemicals with ozone and oxidants in the air. In particular formaldehyde can be formed through these chemical reactions, increasing formaldehyde levels indoors.

> Researchers have begun to understand the potentially larger health threat posed by secondary emission, the chemicals formed by the interactions between oxidants in indoor air and chemicals on surfaces, and by hydrolysis. Many of the by-products of these interactions are more irritating, odorous, or toxic and may pose a far greater health hazard than the chemicals from which they are formed.

Levin warns that introducing outdoor air into buildings may aggravate the problem. Ozone, brought in through ventilation can react with both VOCs and SVOCs to create even more dangerous chemicals such as aldehydes.[18]

VOCs, child health and asthma

There is more evidence of risks from VOCs to babies and young children in the medical literature. In particular VOCs are linked with increased rates of asthma in children.

> Infants and young children have a higher resting metabolic rate and rate of oxygen consumption per unit body weight than adults because they have a larger surface area per unit body weight and because they are growing rapidly. Therefore, their exposure to any air pollutant may be greater. In addition to an increased need for oxygen relative to their size, children have narrower airways than do adults. Thus, irritation caused by air pollution that would produce only a slight response in an adult can result in potentially significant obstruction in the airways of a young child.[19]

The WHO issued guidance in 2008 warning of domestic exposure to VOCs as a cause of asthma in young children.[20] A major literature review of epidemiological studies by Mendell in 2006 shows clear links between a range of indoor air pollutants and an increase in asthma in children. 'The reviewed studies have associated indoor chemical emissions with increased airway inflammation, allergic sensitization, and potentially altered immune system development in infants and children, all processes involved in the genesis and exacerbation of asthma'.[21]

In a study conducted in Perth, Western Australia, of 88 children suffering from asthma, VOC levels were measured in summer and winter in the living room (not bedroom!) of each household. The results were compared with a control group of 104 children. The VOC levels were relatively low compared with 'safe' levels and yet the researchers came to the conclusion that there was a significant correlation with asthma.

> Most of the individual VOCs appeared to be significant risk factors for asthma with the highest odds ratios for benzene followed by ethylbenzene and toluene. For every 10 unit increase in the concentration of toluene and benzene (mg/m^3) the risk of having asthma increased by almost two and three times, respectively. Domestic exposure to VOCs at levels below currently accepted recommendations may increase the risk of childhood asthma. (Measurement of total VOCs may underestimate the risks associated with individual compounds).[22]

The VOCs identified in the homes of participants in this study included benzene, toluene, m-xylene, op-xylene, ethylbenzene, styrene chlorobenzene, 1,3-dichlorobenzene, 1,2-dichlorobenzene and 1,4-dichlorobenzene, with toluene, xylene and benzene + toluene being detected in 99 per cent of the homes studied, xylene in 93 per cent and benzene in 86 per cent.

A study of residents of social housing in southwest England by Exeter University found an increase in reported asthma problems when houses had been renovated to make them more energy efficient. The authors suggested that this might be due in part to 'elevated concentrations of VOCs':

> Energy efficiency may increase the risk of adults reporting that they had visited a doctor in the last 12 months for asthma in this population residing in social housing. A unit increase in household Standard Assessment Procedure (SAP) rating was associated with a 2% increased risk of current asthma, with the greatest risk in homes with SAP N71 … the lack of ventilation and elevated dampness may lead to the proliferation of house dust mites (HDM) and elevated concentrations of volatile organic compounds (VOCs), which are known risk factors for asthma sufferers) and may also explain the association between energy efficiency and asthma.[23]

It is not uncommon for householders to redecorate bedrooms anticipating the arrival of a new baby. This means that the tiny infant is placed in an environment where recent paints and fungicides are off-gassing into the nursery at a much higher rate than normal.[24]

> The first thing I did when we moved in was to change the colour of the nursery. It was a horrible dark blue and I wanted everything clean and white and light and lovely for this pure new life that was about to start. My midwife warned me about the dangers of paint fumes, and as a doctor I am aware of them too, but my hormones seemed to compel me to decorate.
>
> But if you are in the process of choosing between pink, blue or yellow for your new baby's room, maybe you should think twice before picking up the paintbrush. Many paints and varnishes don't just smell funny, they also give off volatile organic compounds (VOCs), which have been associated with birth defects. Water-based paints are less hazardous than solvent-based ones, but even they might contain metal-based pigments. And speaking of metals, sanding down the paintwork in an older house might expose you to a high dose of lead, too.[25]

> Pregnant women told to avoid painting the nursery, buying new furniture … as they may expose their unborn baby to dangerous chemicals. New advice warns to avoid paint fumes, new fabrics, furniture and cars. Guidance is from Royal College of Obstetricians and Gynaecologists.[26]

Not only are VOCs a risk to children but also substances used in water-based paints, known as PGEs (propylene glycol and glycol ether) have been found to be a serious problem. PGE emission from paint is sustained far beyond several months following the paint application. Also PGE emissions increase with higher humidity.

> Children who sleep in bedrooms containing fumes from water-based paints and solvents are two to four times more likely to suffer allergies or asthma, according to a new scientific study. Scientists measured the compounds – propylene glycol and glycol ethers, known as PGEs – in the bedroom air of 400 toddlers and preschoolers, and discovered that the children who breathed them had substantially higher rates of asthma, stuffy noses and eczema.[27]

'Propylene glycol and glycol ether (PGE) in indoor air have recently been associated with asthma and allergies as well as sensitization in children associations of the PGE and the diagnoses of asthma, rhinitis, and eczema, respectively'.[28]

Despite reductions in VOCs due to legislation and the marketing of so-called 'VOC free paints', water-based paints are still seen as a health risk to children.[29] Children breathing fumes from water-based paints have high risk of asthma and allergies, according to a study reported in *Environmental Health News* and a report from the European Environment Agency that confirms the links between VOCs and asthma.

> Domestic exposure to VOCs at levels below currently accepted recommendations may increase the risk of childhood asthma. Measurement of total VOCs may underestimate the risks associated with individual compounds. For every 10 unit increase in the concentration of toluene and benzene (mg/m^3) the risk of having asthma increased by almost two and three times.[30]
>
> While there are no harmonised guidelines for the estimation of total VOCs, values between 300 and 500 μg/m^3 are reported in the literature as an indication of good indoor air quality. That is, however, without any toxicological justification. Domestic exposure to VOCs at lower levels may increase the risk of childhood asthma (6 months–3 years); it has been reported that children exposed to concentrations of VOCs higher than 60 μg/m^3 as the median level of exposure had a fourfold increased risk of having asthma.[31]

Further emissions, having particularly dangerous effects on children come from polyvinyl chloride (PVC)-based flooring materials that contain a phthalate plastic softener. The danger of phthalates in toys and food containers and baby's bottles is well known but these dangerous chemicals are also used in building materials. Di(2-ethylhexyl)phthalate (DEHP) levels in PVC flooring are eight times higher than in alternatives such as linoleum. Manufacturers have lobbied hard, however, not to ban these materials

> Chemical manufacturers say they will seek approval from the European Union to continue use of di(2-ethylhexyl)phthalate (DEHP), a plastic-softening phthalate that the EU is banning.
>
> DEHP is among the first six compounds that the EU is phasing out under its Registration, Evaluation, Authorization & Restriction of Chemical substances (REACH) program. Sale or use of these six chemicals will cease in three to five years unless industry obtains authorization, the European Commission announced on February 17, 2011. In addition to DEHP, the ban affects two other phthalates – benzyl butyl phthalate and dibutyl phthalate. The three phthalates are targeted because of reproductive toxicity. The EU already prohibits use of these three compounds in children's toys.[32]

Official guidelines and standards for VOCs

> The general need for improved source control to diminish the pollution load on the indoor environments from health, comfort, energy efficiency and sustainability points of view leads to the recommendation that VOC levels in indoor air should be kept as low as reasonably achievable (the ALARA principle). ALARA will require that TVOC concentrations in indoor environments do not exceed the typical levels encountered in the building stock of today, unless there are very good and explicit reasons.[33]

Since the EU established the ALARA principle in 1997, many countries and organisations have moved to establish VOC emissions standards and this is explained in more detail below. Some industries such as paint and adhesive manufacturers now market their products as 'low VOC' but this can be somewhat misleading as scientific evidence still points to health problems from so-called low VOC products.

It was always the intention of the EU Construction Product Regulations (EUCPR) to certify health and environmental characteristics.[34] The EUCPR sets out provisions for construction (and other products) to have a Conformité Européene (European Conformity) (CE) mark.

Sadly the CE mark is of little value in that it does not yet provide any guidance as to the environmental performance or health implications of products, even though this was the original intention of the European Union (EU). A bag of cement with a CE mark tells you little more than it is a properly made bag of cement!

The stated intention of the European Commission (EC) is for product groups with an impact on IAQ to take account of and include hygiene and health aspects such as the emission of regulated dangerous substances, when the regulations are next revised (based on mandates issued by the EC in 2013). However rather than setting European standards, limit values for VOCs will be left to national regulations and this means that they are most unlikely to be applied in the UK or Ireland.[35] Thus currently there is almost no restriction or public information on limiting VOC levels. Standards for emissions from products have been agreed (see LCI values below) but applying these so that they can be easily understood by building occupants is a long way from being achieved. The problems related to government and EU policies are discussed further in Chapter 12.

National VOC Limits

National VOC limit regulations are gradually being introduced with Germany, Belgium and France leading the way in this respect. Products in the French market had to be labelled from September 2013 and a simple label showing A plus to C (see Figure 2.3) has to be displayed on products. Table 2.1 shows the VOC limit values for the four classifications in in µg/m³.

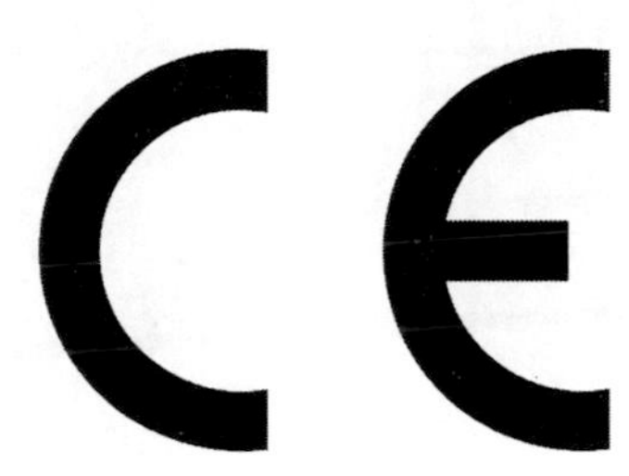

Figure 2.2 The CE mark

Source: www.gov.uk/guidance/ce-marking

Table 2.1 French VOC emissions classes: limit values in $\mu g/m^3$

Classes	*C*	*B*	*A*	*A+*
TVOC	>2000	<2000	<1500	<100
Formaldehyde	>120	<120	<60	<10
Acetaldehyde	>400	<400	<300	<200
Toluene	>600	<600	<450	<300
Tetrachloroethylene	>500	<500	<350	<250
Xylene	>400	<400	<300	<200
1,2,4-trimethylbenzene	>2000	<2000	<1500	<1000
1, 4-dichlorobenzene	>120	<120	<90	<60
Ethylbenzene	>1500	<1500	<1000	<750
2-Butoxyethanol	>2000	<2000	<1500	<1000
Styrene	>500	<500	<350	<250

Germany has set IAQ standards that include VOC levels. A task force of public health authorities (AgBB) and the Deutsches Institut fur Bautechnik (DIBt) developed restrictions for VOCs for a number of construction products, including floor coverings, parquet flooring and adhesives in 2005, targeting 160 toxic VOCs and 30 carcinogens, but the German government has run into trouble trying to enforce standards that go beyond the wholly inadequate provisions of the EUCPR.

A decision of the European Court (C-100/13) clearly said that Germany must not establish a national approval system for construction products on top of CE marking. This was in contradiction with the German approval system and the Ü mark, administered by the DIBt.[36] The EC received numerous complaints from manufacturers and importers who had difficulty in placing their construction products on the German market.[37] This is an indication as to how chemical companies and manufacturers can use European harmonization legislation to block any country from raising indoor air quality standards.

Belgium has established a TVOC limit of 1000 $\mu g/m^3$ with 200 $\mu g/m^3$ for acetaldehyde, 300 $\mu g/m^3$ for toluene and 100 $\mu g/m^3$ for formaldehyde. Work is proceeding to establish overall EU values[38] and the European Committee for Standardization (CEN)[39] has published VOC testing standards CEN/TS 16516 – and there is an ISO[40] standard for testing VOCs (ISO 17025). The issue surrounding what are being called lowest concentrations of interest (LCI) values is discussed further in a later chapter.

Figure 2.3 French emission label

Source: www.eco-institut.de/en/certifications-services/international-labelling/french-voc-label/

In the UK, ventilation standards in the UK building regulations (Approved Document F) includes VOC limit guidance: 'Total Volatile Organic compound (TVOC) levels should not exceed 300 μg/m^3 averaged over 8 hours'.[41] However such a standard, as far as can be found, is rarely, if ever, enforced, and most professionals involved in the UK construction industry are unlikely to be aware of it or understand what it means. In any case a scientific view is emerging that TVOCs are not an effective way of setting VOC standards as they are not a sufficiently rigorous method for limiting hazards.

VOC content levels for paints were established in UK law in 2012.[42] As a result VOC emissions from paints have been significantly, but not completely, reduced. In EU countries where there are no VOC emissions regulations (such as the UK) CE marked products can be sold with a 'No Performance' declaration. Products with a low or even no VOC emissions into indoor air can be subject to CE marking on basis of a 'without testing' (WT) option or a 'without further testing' (WFT) option.[43]

There has been a head-in-the-sand attitude to VOCs in the UK and for many years official guidance claimed that VOCs were not a risk to health.

> According to the Department of Health Committee on the Medical Effects of Air Pollutants, there is no evidence to suggest that current UK domestic exposures to VOCs, either as individual chemicals or as a total, pose a risk to health. The contribution of VOCs towards carcinogenic, mutagenic risk and neurotoxic effects in the UK population is considered negligible.[44]

A document was prepared by the Institute of Environmental Health (IEV) in 1999 entitled *Volatile organic compounds (including formaldehyde) in the home* and was intended to provide advice to environmental health officers but the IEH quoted the above statement that VOC risk is negligible. However they did spell out the risks as follows:

> There have been reports of exposure to VOCs that lead to sensory effects and general discomfort due to strong odours, and irritancy or allergic reactions. At extremely high levels of exposure headaches, eye and throat irritation, nausea, dizziness and drowsiness are some of the symptoms that may be experienced. Individuals with existing respiratory diseases, such as asthma or allergy, may be particularly susceptible and may react to VOC exposure at levels below those that would affect healthy individuals. A few VOCs are known to pose health risks at lower exposures
>
> A small number of VOCs found at low concentrations in indoor air are recognized carcinogens. Some, such as benzene, are genotoxic and no absolutely safe level of exposure can be defined. However, at the concentrations measured in most of the homes studied, the risk to health from these compounds is negligible.[45]

Postnote, a briefing document for UK members of Parliament produced in 2010, has drawn attention to health risks from VOCs and indicates how awareness of VOC risks has grown.

> VOCs include both natural and synthetic chemicals, e.g. formaldehyde, from a variety of sources such as construction products, cleaning products, air fresheners, paints and electrical goods. Research suggests VOCs can irritate the lungs, particularly in children. VOCs may also react with ozone (an outdoor air pollutant known to cause respiratory inflammation produced indoors by some printers) to produce other toxic compounds.[46]

For many years BRE's environmental assessment tool, BREEAM, largely ignored IAQ and health issues and continued to give 'A plus' ratings to products with serious health and emission concerns. However BREEAM recently introduced VOC testing limits as part of the BREEAM scoring system (Hea 02 Indoor Air Quality). Three credits are available for minimising sources of air pollution:

> *An indoor air quality plan has been produced which considers:*
> *a. Removal of contaminant sources*
> *b. Dilution and control of contaminant sources*
> *c. Procedures for pre-occupancy flush out*
> *d. 3rd party testing and analysis.*

BRE was contacted and asked for evidence of projects where health/VOC points had been awarded but they refused to provide any information.

> Please accept our apologies for the delay (5 weeks) in responding to this query. Unfortunately we are unable to provide details on specific projects and how and if they have achieved certain credits in BREEAM, as the information itself is confidential and we would require assessor confirmation in order to release any project specific information.[48]

BREEAM Hea 02 Criteria
One credit
Criterion 1 is achieved.
All decorative paints and varnishes have met the requirements listed in Table – 8
At least five of the eight remaining product categories listed in Table – 8 have met the testing requirements and emission levels for Volatile Organic Compound (VOC) emissions against the relevant standards identified within this table. Where five or less products are specified within the building, all must meet the requirements in order to achieve this credit.
One credit
9. Criterion 1 is achieved
10. Formaldehyde concentration level is measured post construction (but pre-occupancy) and is found to be less than or equal to 100μg/m^3 averaged over 30 minutes (WHO guidelines, source BRE Digest 464 part 220).
11. The total volatile organic compound (TVOC) concentration is measured post construction (but pre-occupancy) and found to be less than 300μg/m^3 over 8 hours, in line with the Building Regulation requirements.
12. Where levels are found to exceed these limits, the project team confirms the measures that have, or will be undertaken in accordance with the IAQ plan, to reduce the TVOC and formaldehyde levels to within the above limits.
14. The measured concentration levels of formaldehyde (μg/m3) and TVOC (μg/m^3) are reported, via the BREEAM scoring and reporting tool, for the purpose of confirming criteria 10 to 12.

Figure 2.4 Extract from the BREEAM Hea 02 criteria
Source: BREEAM[47]

A helpful BREEAM assessor was able to extract the information from the BRE and was told that out of a random sample of 550 certified projects, 240 achieved a Hea 02 credit, so more than half appear to be considering the new health criteria despite the fact that far fewer credits are available than for other issues such as energy efficiency. Hea 02 takes up 13 pages of the BREEAM guide and would require a good understanding of IAQ issues and testing methodologies, so BREEAM assessors may require the help of IAQ specialists.

A great deal of work has been carried out on VOC dangers in the USA by a wide range of organisations including the US EPA.

> EPA's Office of Research and Development's 'Total Exposure Assessment Methodology (TEAM) Study' found levels of about a dozen common organic pollutants to be 2 to 5 times higher inside homes than outside, regardless of whether the homes were located in rural or highly industrial areas. TEAM studies indicated that while people are using products containing organic chemicals, they can expose themselves and others to very high pollutant levels, and elevated concentrations can persist in the air long after the activity is completed. Reducing the concentration of VOCs indoors and outdoors is an important health and environmental goal. For indoor air quality, ALL organic chemical compounds whose compositions give them the potential to evaporate under normal atmospheric conditions are considered VOCs and should be considered in any assessment of indoor air quality impacts.[49]

> VOCs cause eye, nose, and throat irritation, headaches, nausea, and can damage the liver, kidney, and central nervous system. [They] are chemicals that evaporate at room temperature. VOCs are emitted by a wide array of products used in homes including paints and lacquers, paint strippers, varnishes, cleaning supplies, air fresheners, pesticides, building materials, and furnishings. VOCs are released from products into the home both during use and while stored.[50]

The US State of California Health and Safety Code states that no-one should discharge from any source whatsoever (including VOCs) contaminants that could endanger the health of the public and methods for measuring VOC emissions have been adopted. It was not possible to find any examples of penalties for infringement of these standards however (see Figure 2.5).[51]

WARNING
This Area Contains
A Chemical
Known To The
State Of California
To Cause Birth Defects
Or Other Reproductive Harm.
California Health & Safety Code Section 25249.5

Figure 2.5 California Health and Safety Code warning notice

VOC emissions from products

Building materials, furnishings, paints, office equipment and consumer products are significant sources of VOCs in buildings, motor vehicles, aircraft and other environments. BRE says that they can have an adverse effect on health and comfort, including eye and airway irritation, headaches and tiredness.[52] BRE offers air quality surveys and emission testing[53] but so do many other organisations.

What are the risks from VOCs?

WHO has produced an excellent 450-page document that provides guidelines for the protection of public health from risks due to chemicals commonly present in indoor air. The document analyses the following:

- benzene
- carbon monoxide
- formaldehyde
- naphthalene
- nitrogen dioxide
- polycyclic aromatic hydrocarbons
- radon
- trichloroethylene
- tetrachloroethylene

All these have indoor sources and are known as hazardous to health and are often found indoors in concentrations of health concern.

> The primary aim of these guidelines is to provide a uniform basis for the protection of public health from adverse effects of indoor exposure to air pollution, and to eliminate or reduce to a minimum exposure to those pollutants that are known or are likely to be hazardous.[54]

While there are numerous scientific papers that provide strong evidence of high VOC emissions in indoor air and even more on methods for measuring this, there is insufficient data on what effect they might have on occupants.

VOC emissions are likely to be at their highest in new buildings or buildings that have been recently refurbished or decorated. Hotel rooms and corridors commonly smell of fresh paint and new carpet as they undergo continuous 'refreshing'. A common view has been that high-level emissions dissipate in the first few weeks to a low level, which no longer presents a hazard. This assumption may be incorrect, in that new buildings are rarely properly ventilated, and VOCs may linger for much longer than a few weeks.

> Poor indoor air quality can partly be attributed to emissions of formaldehyde and volatile organic compounds (VOCs) from building materials. These emissions negatively affect people's comfort, health, and productivity. Formaldehyde, which is regarded as a human carcinogen, is of particular concern in the indoor environment.[55]

In an indoor air quality study of a large single family house in Finland in 2014 researchers from East Finland University found high concentrations of VOCs when the occupants moved in and even when a ventilation system had been turned on, though these reduced over an eight-month period.

> VOCs were mainly emitted from the wood based materials (terpenes), sealants and water based paints (oximes and alcohols). Some of the identified compounds (e.g. 2-propanone oxime, 2-butanone oxime) found are strong sensory irritants. Concentrations of VOCs were at high levels for several weeks after the inhabitants moved into the new house. Especially concentrations of terpenes (mainly alpha-pinene and delta-carene) stayed a high level. The results revealed that harmful VOCs from finishing materials and furnishings are present in brand new houses, indicating that efficient ventilation should be used before the inhabitants move in.[56]

VOCs react with ozone to create secondary emissions hazards

Ozone from external air can react with VOCs in indoor air to create secondary emissions. As long ago as 1992 Weschler reported experiments with ozone and carpets. He showed that the emissions from carpet of styrenes and of 4-phenylcyclohexene (4-PCH), a by-product of the styrene butadiene rubber (SBR) latex manufacturing process, would rapidly react with ozone indoors to form formaldehyde, acetaldehyde, and other, higher molecular weight aldehydes. SBR latex backings were the most common on commercial carpet at that time and represented about 85 per cent of the commercial carpet market. Numerous further studies have confirmed similar reactions between ozone and VOCs.[57,58]

Levin has warned of the dangers from secondary emissions, the chemicals formed by the interactions between oxidants in indoor air and chemicals on surfaces, and by hydrolysis. Levin, Weschler and others point out that activated carbon can be used in air conditioning and mechanical ventilation systems to remove ozone but this is unlikely to be included in domestic systems so the risk of secondary emissions is therefore higher.[59,60]

PVC flooring is widely used, particularly in hospitals and school buildings but has been identified as a source of VOC emissions and even more dangerous through ozone reacting with the PVC materials. 'Ozone deposition is considered as a surface reaction between ozone and unsaturated organics on a PVC material surface. These reactions produce pollutants that can be released into the gas-phase as secondary emissions'.[61] The phthalates used in PVC are dangerous for health, affecting asthma but also act as an endocrine disruptor.

> Phthalates are known carcinogens and have been identified as likely asthmagens. These additives are not tightly bound to the PVC molecules and are known to migrate from PVC products. They are commonly found in both environmental dust studies in buildings with PVC products and in biomonitoring studies of occupants.[62]

Building materials as significant sources of VOCs

Wood and composite wood products

Wood composite products are a significant source of VOCs because of the adhesives and resins that are used. Chipboard, oriented strandboard (OSB), medium-density fibreboard (MDF), particleboard, hardboard, plywood and laminated products are used in significant quantities in nearly all buildings. The main cause of concern is that of formaldehyde emissions that have been a mainstay of resin adhesives, however as numerous countries have imposed formaldehyde emission levels on composite wood products, other glues have been used such as PFA (professional flooring adhesive) and amino plastics and phenolics. Many kitchen units are now plastic coated and there may be emissions from the plastic materials used as well as the adhesives.

In one study, hardboard was found to be the highest emitter of VOCs when compared with other boards, but emission levels between different products vary enormously depending on the glues used.

> In this study the measured maximum TVOCs concentrations from wood products were from 18–408μg/m^3. These values were much lower than indoor concentrations in residences and buildings. In fact, indoor concentrations of TVOCs are the outcome of TVOCs emissions from many other materials and household products.[63]

It is difficult to identify the different VOCs emitted from composite wood products because most scientific studies refer to TVOC emissions, however some studies identify toluene, xylene, 2-butanol, ethyl acetate, acetone, ethanol and alkanes as VOC emissions from wood products.[64]

Heat-treated wood has been identified as a source of unacceptable odours though it has been suggested that this can be reduced by the use of sealants. However it would be unusual to use heat-treated wood internally. 'It is possible to reduce the unwanted odour from heat-treated wood and thus to increase the potential for indoor use. Recently, an international company for interior design stopped the sale of traditional heat-treated wood for indoor use as a result of its odour'.[65] Thermally treated wood has a characteristic odour, which is caused by changes that occur during heat treatment but the TVOC values are claimed to be significantly lower than those for untreated softwoods due to the evaporation of most of the terpenes during the heat treatment process.[66,67]

The VOCs that can be emitted by flooring materials and adhesives include:[68]

- acetic
- 2-ethyl-1-hexanol
- 4-phenylcyclohexan
- hexamethylcyclotrisiloaxne
- benzothisazole
- ethandiol
- diethylenglykol
- 2-methyl—4-isothiazolin-3-on
- benzaldehyd
- cyclohexanon
- n-buthylether
- 1,4-butandiol
- n-methyl-2-pyrrolidon
- diethylenglykol-monobutylether
- longifolen
- acetone
- napthaline
- 1-butylacetate
- a-pinen

- εε pinen
- limonen
- 1-butanol

Paints, stains, varnishes and finishes

Odours and VOC emissions from paints, stains and varnishes would be the most commonly understood by the general public. Decorating buildings gives rise to emissions from the solvents and carriers used in a wide range of products and while the strong smells emitted can dissipate quite quickly, VOC emissions may remain within the building for some considerable time.

The EU has been active in trying to reduce emissions of VOCs from paints and the Paints Directive (PD or DECO 2004/42/EC) of 2004 has set limit values of emissions from paints, which had to be implemented by 2007 and 2010. This was incorporated into UK legislation.[69]

The Solvent Emission Directive (SED) 1999/13/EC was implemented in 2007. This sets down limits on VOC emissions but it is largely concerned with manufacturing and workplace problems ranging from car spraying to dry cleaning. While it may have had some impact on reducing the use of solvents and VOCs, this does not mean that VOC emissions from materials used in buildings will have significantly fallen.[70] The EU PD introduced VOC limits for specific products and materials such as paints that contain VOCs with the aim of preventing the negative environmental effects of emissions of VOCs from decorative paints and vehicle refinishing products.[71] While water-based paints still pose health risks, VOC emissions from solvent-based paints are much higher.

The principle source of the VOC problem with paints was the solvent content derived from petrochemical materials. These include aliphatic hydrocarbons, mineral spirits, naptha, ethyl alcohol, xylene and toluene. Some of these are known as aromatic solvents and all of these chemicals can be damaging to health and IAQ. Solvent-based paints are sometimes referred to as oil or alkyd based. As the solvents evaporate during the application of paints, the chemicals are emitted into indoor air. Paint thinners and brush cleaners also contain solvents and can be the cause of VOC emissions in buildings.

Due to the EU directive, the paint industry has largely switched to water-based paints, referred to as acrylic emulsions. However it is important to realise that many paint products advertised as 'solvent free' may still contain some solvent within the permitted limits and that many of the other materials used in emulsions are also bad for indoor air. Acrylic paints are still synthetic and made from petrochemical-based polymers. They should not be confused with natural or eco paints, though some eco paints may also contain acrylics and even solvents.

Table 2.2 VOC emissions from coatings – an example

Product subcategory	*Type*	*Grams per litre of VOC ready to use*
One pack performance coatings	Water based	140
	Solvent based	500
Two pack reactive coatings for uses such as floors	Water based	140
	Solvent based	500

Source: Based on *A short Introduction to VOC Directives* produced by a coatings manufacturer[72]

Table 2.3 Maximum VOC content for paints from EU directive

Maximum VOC content for paints and varnishes from EU Decopaint Directive 2004/42/EC. Standards apply from 1.1.2010, based on information from Eurofins

Extracts	*Examples*		*g/l*
1	Interior matt walls and ceilings	Water based	30
		Solvent based	30
2	Interior gloss walls and ceilings	Water based	100
		Solvent based	100
5	Interior trim varnishes and wood stains	Water based	130
		Solvent based	300
9	One pack performance coatings	Water based	140
		Solvent based	500

Source: ec.europa.eu/environment/air/pollutants/.../paints/paints_legis.htm

To reduce risks from VOC emissions, the simplest solution is to avoid the use of all synthetic paints but this is not always a simple task when paint manufacturers do not provide a full declaration of the contents of their products. A number of paints contain antimicrobial and fungicidal chemicals but inspection of product data sheets from paint manufacturer provides very little information as to toxicity or hazards simply by stating 'No Information'.

Some natural paints are made from natural oils such as linseed and natural turpentine, and some ecological paints fell foul of the EU paints directive as even natural oils were regarded as emitting VOCs. However there are still natural paints that use linseed oil, some claiming that the linseed is from organic sources. Additives can be natural using chalk and clay and they can be natural mineral based but some clay-based paints may contain polyvinyl acetate and methylcellulose.

Mineral-based paints are made from potassium silicate or sodium silicate. By and large breathable or micro-porous paints are preferable in terms of IAQ as they will help with reducing the risks of mould and damp, whereas acrylic paints will seal up walls and prevent them from breathing.

A number of large commercial paint manufacturers advertise water-based acrylic emulsion paints as healthy and contributing to good IAQ, using terms such as 'breatheasy' but these claims should be treated with some scepticism. Consumers have complained of bad odours from 'odour free paints'.[73]

> Many of the terms used for marketing paint can be confusing – organic paint, for instance, is simply paint which contains carbon compounds and many of the water based gloss paints being marketed as 'environmentally friendly' contain more chemicals than the oil based paints they replace. Low odour paints may be more pleasant to use but just because you can't smell the fumes doesn't mean that they are not still present.[74]

Insulation and general building materials and VOCs

Manufacturers of insulation products tend to claim that their products are not responsible for VOCs and IAQ problems.'"It's pretty clear in the industry that fiberglass insulation is not a contributor to indoor-air quality issues," says Gale Tedhams, manager for Owens Corning residential insulation products'.[75] In some forms of building construction, insulation materials are kept well away from the indoor environment as they are contained within masonry cavities or other non-breathable materials. However increasingly insulation materials are only a few millimetres away from the indoor environment behind vapour permeable plasterboards (dry wall) in lofts and walls. In recent years there has been increasing use of spray polyurethane insulation onto ceilings and in walls, in some cases, without any layer in between the insulation and the indoor environment.

VOC emissions from insulation products and the chemicals, used to make insulations, are found in IAQ tests when there is no other possible source. Chemicals emitted include blowing agents; flame-retardants and other constituents of synthetic insulation products have been found in surprisingly high concentrations in blood tests. As a result, countries are setting VOC limits and controls that include insulation products, such as France.[76]

In the USA the Leadership in Energy and Environmental Design (LEED) rating system now includes emissions from insulation materials.

> Acoustic ceilings, wall panels, wall coverings, textiles, and thermal insulation are new categories that are being incorporated within the Indoor Environmental Quality (IEQ) scopes of green building rating systems, codes, and purchasing specifications. The LEED v4 EQ Credit: Low-Emitting Materials emphasizes a holistic approach where at least 50% of all materials including all layers comprising a wall, floor, or ceiling system including insulation need to be compliant with VOC emission requirements in order for a building project to earn credit. Federal weatherization programs for homes are sensitive to potential IAQ issues and are now focusing on the VOC emissions of thermal insulation.[77]

Some insulation products contain formaldehyde and insulation materials that contain fibres such as fiberglass can irritate the skin, eyes and respiratory tract when disbursed in the air and/or inhaled.[78]

> A major indoor air pollutant of concern is formaldehyde. A large source of this VOC is in pressed wood products, as well as fiberglass insulation … Other indoor air pollutants include other VOCs such as tetrachloroethylene, trichloroethylene, chloroform, benzene, styrene, and so on. When choosing fiberglass insulation, it is best to go with a formaldehyde free brand. There are companies that offer formaldehyde free insulation … The extreme concern of indoor pollution is that on average, people spend 90 percent of their time indoors. Studies show that human exposure to air pollutants suggest that indoor levels of many air pollutants may be two to five times (and sometimes much higher) more than outdoor levels.[79]

Because of concerns about formaldehyde in fibreglass and other mineral fibre insulations, some manufacturers have developed new products to reduce the formaldehyde content, as discussed by the USA Pharos project in relation to Knauf Ecose insulation.

> The Knauf EcoBatt products use the Ecose binder, for which we only found orange flagged or better chemicals in the patents. They are, however, reportedly making the insulation on the same machines as their other formaldehyde-based products and so Knauf acknowledges that Ecose products may have trace amounts of urea phenol formaldehyde binder. This should result in far less formaldehyde releases than a UPF-bound product,

> but it means the products can't pass the Pharos formaldehyde-free filter until they complete the transition. Not only is the binder formaldehyde-free, but it is also partially plant-based, However, since the binder makes up 17% or less of the overall product content and the plant component is only 20% of the binder, the total bio-based content (about 3%) is too small to affect renewable material scoring.[80]

While most attention has focused on formaldehyde, insulation materials may emit a whole range of other VOCs and hazardous gases, though the synthetic insulation industry is very active in denying the risks. For example the US Polyurethane foam association states 'Bottom-line, flexible polyurethane foam does not typically emit significant amounts of either EPA-listed VOCs or "semi-VOCs." (With the understanding that with today's extremely sensitive testing capabilities – there may be no such thing as a true "zero.")'.[81]

Styrenes are of particular concern as they are widely used in insulation and other construction products. Styrenes are emitted from cigarette smoke and are regarded as most hazardous when combined with emissions from building materials, as reported by the Agency for Toxic Substances and Diseases Registry in the USA.

> The principal route of styrene exposure for the general population is probably by inhalation of contaminated indoor air. Mean indoor air levels of styrene have been reported in the range of 0.1–50 $\mu g/m^3$, and can be attributed to emissions from building materials, consumer products, and tobacco smoke.[82]

VOCs are emitted by building materials outside and inside buildings. They are a risk to adult health and particularly child health. Moves are being made to establish standards and limits and ways of assessing VOC levels in buildings and within a relatively short period of time. Assessing and reducing VOCs will become standard practice in the design, construction and commissioning of buildings. However in order to understand why we fill buildings with such hazardous materials we need to ask why we have to use so many chemicals in the construction industry and society at large.

Notes

1 Volatile Organic Compounds' Impact on Indoor Air Quality, Environmental Protection Agency (EPA) www.epa.gov/indoor-air-quality-iaq/volatile-organic-compounds-impact-indoor-air-quality [viewed 28.1.16]

2 Crump. D. (2011) Nature and Sources of Indoor Chemical Contaminants. In Guardino Solá (ed.), *Encyclopaedia of Occupational Health and Safety*, International Labour Organization

3 From a typical home survey report by Prism Analytical Technologies Waverton Analytics 2015 www.homeaircheck.co.uk [viewed 28.1.16]

4 Hess-Kosa K. (2011) *Indoor Air Quality: The Latest Sampling and Analytical Methods,* Second Edition, CRC Press

5 WHO Europe (2011) *Methods for Monitoring Indoor Air Quality in Schools*. European Commission

6 Bacaloni A., Insogna, S. and Zoccolillo, L. (2011) Indoor Air Quality: Volatile Organic Compounds: Sources, Sampling and Analysis. In Mazzeo, N. (ed.) *Environmental Sciences Chemistry, Emission Control, Radioactive Pollution and Indoor Air Quality.* INTECH

7 Lee, C.S., Haghighat, F. and Ghaly, W.S. (2005) A study on VOC source and sink behavior in porous building materials: analytical model development and assessment. *Indoor Air* June 15 (3), 183–196.

8 Salthammer, T. (2009) *Organic Indoor Air Pollutants: Occurence, Measurement, Evaluation*. Second Edition. Basingstoke: Wiley

9 Spiegel, R. and Meadows, D. (2010) *Green Building Materials: A Guide to Product selection and specification*. Wiley

10 Molhave, L., Back, B. and Pedersen, O.F. (1986) Human reactions to low concentrations of volatile organic compounds, *Environment International* 12, 167–175

11 Kjaergaard, S.K., Molhave, L. and Pedersen, O.F. (1991) Human reactions to a mixture of indoor air volatile organic compounds, *Atmospheric Environment* 25, 1417–1426

12 Fiedler, N., Laumbach, R., Kelly-McNeil, K., Lioy, P., Fan, Z.H., Zhang, J., Ottenweller, J., Ohman-Strickland, P. and Kipen, H. (2005) Health effects of a mixture of indoor air volatile organics, their ozone oxidation products, and stress, *Environmental Health Perspectives* 113 (11), 1542–1548

13 Hutter, H.P., Moshammer, H., Wallner, P., Tappler, P. and Kundi, M. (2005) *Volatile Organic Compounds: guidelines from the Austrian working group on indoor air*, Medical University Vienna, Medicine and Environmental Protection, Danube University Krems

14 Sundell, J. (2004) On the history of indoor air quality and health, *Indoor Air*, 14 (7), 51–58

15 Zhang, J. and Smith, K.R. (2003) Indoor air pollution: a global health concern, *British Medical Bulletin, Oxford Journals Medicine & Health* 68 (1), 209–225

16 Black, M.S. (2015) Reducing the risk of VOC emissions: a product emissions review, Underwriters Laboratories Inc., Marietta, GA, Healthy Buildings Europe Conference, Eindhoven May 2015

17 O'Neal-Coble, L., Holland & Knight LLP (2011) *Indoor Air Quality Problems in Buildings in the United States*, Holland & Knight LLP www.academia.edu/2507901/indoor_air_quality_problems_in_buildings_in_the_united_states [viewed 20.3.17]

18 Levin, H. (2008) The big indoor air emissions threat – secondary emissions, World Sustainable Buildings Conference, September, Melbourne

19 Pronczuk-Garbino, J. (2005) *Children's Health and the Environment – A global perspective. A resource guide for the health sector*, World Health Organization

20 World Health Organization (2008) *Children's Health and the Environment, WHO Training Package for the Health Sector*, Geneva: WHO www.who.int/ceh [viewed 2.2.16]

21 Mendell, M.J. (2007) Indoor residential chemical emissions as risk factors for respiratory and allergic effects in children: a review, *Indoor Air Journal* 17, 259–327

22 Rumchev, K., Spickett, J., Bulsara, M., Phillips, M. and Stick, M. (2004) Association of domestic exposure to volatile organic compounds with asthma in young children, *Thorax* 59 (9), 746–751

23 Sharpe, R.A., Thornton, C.R., Nikolaou, V. and Osborne, N.J. (2015) Higher energy efficient homes are associated with increased risk of doctor diagnosed asthma in a UK subpopulation, *Environment International* 75, 234–244

24 Parents' tips: decorating your nursery on a budget, Baby-centre www.babycentre.co.uk/a25004939/parents-tips-decorating-your-nursery-on-a-budget [viewed 2.2.16]

25 Hesse, K. (2006) Step away from the paint roller, *The Telegraph* www.telegraph.co.uk/finance/property/3349062/Step-away-from-the-paint-roller.html [viewed 2.2.16]

26 Hope, J. (2013) Pregnant women told to avoid painting the nursery, buying new furniture or going near non-stick frying pans as they may expose their unborn baby to dangerous chemicals, www.dailymail.co.uk/health/article-2336030/A-mum-Dont-paint-nursery-avoid-non-stick-frying-pans-Pregnant-women-warned-risk-baby-exposure-chemicals.html#ixzz3l2jjM5B6 [viewed 2.2.16]

27 Cone, M. (2010) Volatile organic compounds may worsen allergies and asthma, *Scientific American*, www.scientificamerican.com/article/volatile-organic-compounds [viewed 2.2.16]

28 Choi, H., Schmidbauer, N., Spengler, J. and Bornehag, C.G. (2010) Sources of propylene glycol and glycol ethers in air at home, *International Journal of Environmental Research and Public Health* 7 (12), 4213–4237

29 Goetzman, K. (2010) Water-based paints pose health risk to kids, www.utne.com/environment/water-based-paints-pose-health-risk-to-kids.aspx [viewed 2.2.16]
30 European Environment Agency (2013) Environment and human health, Joint EEA-JRC report, EEA Report No 5/2013 Report EUR 25933 EN
31 Rumchev K., *et al.*, op cit. (2004)
32 Europe calls this one of the top 6 chemical threats to humans, MMS Healthy for Life www.mmshealthyforlife.com/europe-calls-this-one-of-the-top-6-chemical-threats-to-humans [viewed 2.2.16]
33 European Commission Joint Research Centre – Environment Institute (1997) *TVOCs in Indoor Air Quality Investigations, European Collaborative Action Indoor Air Quality and its Impact on Man*, Report No. 19, EC
34 VOC emissions under construction products regulation, Eurofins www.product-testing.eurofins.com/information/compliance-with-law/european-directives-and-laws/construction-products/voc-emissions-under-cpr.aspx [viewed 2.2.16]
35 CP-DS: Legislation on substances in construction products, European Commission http://ec.europa.eu/growth/tools-databases/cp-ds/ [viewed 3.2.16]
36 Will the German DIBt approval system have a future?, Eurofins https://voctesting.wordpress.com/2015/04/15/will-the-german-dibt-approval-system-have-a-future/ [viewed 3.2.16]
37 Gates K&L (2015) European Court of Justice overturns additional requirements for the marketing of construction products in German Building Rules List, Construction Law www.klconstruction-lawblog.com/2015/01/articles/articles-and-publications/european-court-of-justice-overturns-additional-requirements-for-the-marketing-of-construction-products-in-german-building-rules-list/ [viewed 3.2.16]
38 Lists of substances and EU-LCI values www.eu-lci.org/EU-LCI_Website/EU-LCI_Values.html [viewed 3.2.16]
39 CEN www.cen.eu [viewed 3.2.16]
40 ISO www.iso.org/iso/home.html [viewed 3.2.16]
41 www.planningportal.gov.uk/uploads/br/BR_PDF_ADF_2006.pdf [viewed 15.3.16]
42 The Volatile Organic Compounds in Paints, Varnishes and Vehicle Refinishing Products Regulations 2012, deddfwriaeth.gov.uk www.legislation.gov.uk/cy/uksi/2012/1715/made [viewed 3.2.16]
43 VOC Emissions under Construction Products Regulation, Eurofins www.eurofins.com/product-testing-services/information/compliance-with-law/european-directives-and-laws/construction-products/voc-emissions-under-cpr.aspx [viewed 3.2.16]
44 MRC Institute for Environment and Health (1999) Volatile organic compounds (including formaldehyde) in the home www.iehconsulting.co.uk/IEH_Consulting/IEHCPubs/AirPollution/volatile-organic-compounds-including-formaldehyde-in-the-home.pdf [viewed 3.2.16]
45 www.iehconsulting.co.uk/IEH_Consulting/IEHCPubs/AirPollution/volatile-organic-compounds-including-formaldehyde-in-the-home.pdf [viewed 4.2.16]
46 Postnote 366, November 2010 www.parliament.uk/documents/post/postpn366_indoor_air_quality.pdf [viewed 4.2.16]
47 BREEAM (2011) *BREEAM New Construction, Non Domestic Buildings, Technical Manual SD5073-2.O.2011*, London: BRE
48 Email from BREEAM Technical Consultant Charlotte Hardy 7.7.2015
49 www.epa.gov/indoor-air-quality-iaq/volatile-organic-compounds-impact-indoor-air-quality [viewed 6.2.16]
50 United States Environmental Protection Agency (2008) *Care for Your Air: A Guide to Indoor Air Quality*, EPA 402/F-08/008 www.epa.gov/indoor-air-quality-iaq/care-your-air-guide-indoor-air-quality [viewed 3.2.16]
51 www.leginfo.ca.gov/cgi-bin/displaycode?section=hsc&group=41001-42000&file=41700-41712 [viewed 14.3.16]
52 VOC emissions from building and consumer products, BRE www.bre.co.uk/filelibrary/pdf/cap/KN2567_VOC_Emissions_Cap_Sheet.pdf undated [viewed 25.2.16]

53 www.bre.co.uk/page.jsp?id=1417 [viewed 25.2.16]

54 WHO (2010) *Guidelines for Indoor Air Quality: selected pollutants*. Geneva: WHO

55 Xiong, J., Yao, Y. and Zhang, Y (2011) C-History method: rapid measurement of the initial emittable concentration, diffusion and partition coefficients for formaldehyde and VOCs in building materials, *Environmental Science and Technology* 45, 3584–3590, 10.1021/es 200277p

56 Hyttinen, M. and Pasanen, P. (2015) Volatile organic compounds during and after the construction in new built single-family house, *Healthy Buildings Europe conference proceedings*, Eindhoven

57 Weschler, C.J. and Shields, H.C. (1999) Indoor ozone/terpene reactions as a source of indoor particles *Atmospheric Environment* 33 (15), 2301–2312

58 Weschler, C.J. (2006) Ozone's impact on public health: contributions from indoor exposures to ozone and products of ozone-initiated chemistry, *Environ Health Perspect*. 114 (10), 1489–1496.

59 Levin, H. (2008) The big indoor air emissions threat – secondary emissions. World Sustainable Buildings Conference, SB08 Melbourne, September 21–25

60 Sidheswaran, M.A., Destaillats, H., Sullivan, D.P., Cohn, S. and Fisk, W.J. (2011) Energy efficient indoor VOC air cleaning with activated carbon fiber (ACF) filters *Building and Environment xxx*, 1–11

61 Ward, M.K.M., Mendez, M. and Schoemaecker, C. (2015) VOC emissions from ozone initiated surface reactions with PVC flooring from a classroom. *Healthy Buildings Europe*, Eindhoven

62 Valette, J. *et al.* (2015) Post-consumer polyvinyl chloride in building products: A Healthy Building Network evaluation for Stop Waste and the Optimizing Recycling Collaboration. *Healthy Building Network*, April

63 Guo, H., Murraya, F. and Leeb S.-C. (2002) Emissions of total volatile organic compounds from pressed wood products in an environmental chamber, *Building and Environment* 37, 1117–1126

64 Yrieix, C., Maupetit, F. and Ramalho, O. (2004) Determination of VOC emissions from French wood products. 4th European Wood-Based Panel Symposium, September, Hanover

65 Sandberg, D., Haller, P. and Navi, P. (2013) Thermo-hydro and thermo-hydro-mechanical wood processing: An opportunity for future environmentally friendly wood products, *Wood Material Science & Engineering* 8 (1), (no page nos)

66 Key properties of thermowood, MestaWood www.metsawood.com/global/Products/thermowood/Pages/Key-properties.aspx [viewed 3.2.16]

67 Hyttinen, M., Masalin-Weijo, M.M., Kalliokoski, P. and Pasanen, P. (2010) Comparison of VOC emissions between air-dried and heat-treated Norway spruce (Picea abies), Scots pine (Pinus sylvesteris) and European Aspen (Populus tremula) wood. *Atmospheric Environment* 44. 5028–5033

68 Müller, B., Panašková, J., Danielak, M., Horn, W., Jann, O. and Müller, D. (2011) BAM report, Sensory based evaluation of products emissions, *Umwelt Bundesamt* 61 www.umweltbundesamt.de/en/publikationen/sensory-based-evaluation-of-building-product [viewed 3.2.16]

69 http://ec.europa.eu/environment/air/pollutants/stationary/paints/paints_legis.htm [viewed 13.3. 16]

70 DEFRA (2010) *Environmental Permitting Guidance: The Solvent Emissions Directive For the Environmental Permitting (England and Wales) Regulations March 2010,* Department of Food and Rural Affairs

71 http://ec.europa.eu/environment/air/pollutants/stationary/paints/paints_legis.htm [viewed 2.2.16]

72 www.hempel.co.uk/~/media/Sites/hempel-co.../hempel-**voc**-gb.pdf [viewed 2.2.16]

73 http://whatconsumer.co.uk/forum/consumer-rights-television-programmes/10608-bbc-watchdog-crown-paint-not-sniffed.html [viewed 2.2.16]

74 https://decoratingadvice.co.uk/paint/eco-paints [viewed 2.2.16]

75 Miller, S., Fiberglass insulation, Remodelling www.remodeling.hw.net/products/exteriors/fiberglass-insulation [viewed 3.2.16]

76 www.eurofins.com/media/3686245/French%20VOC%20Regulations%20-%20ch-en.pdf [viewed 2.2.16]

77 Test your interior building products for VOC emissions, Berkeley Analytical http://berkeleyanalytical.com/industries/building-products/ [viewed 3.2.16]

78 VOCs in fiberglass insulation, Appropedia www.appropedia.org/VOCs_in_fiberglass_insulation [viewed 3.2.16]

79 Lent, T. (2010) New EPA study confirms health dangers of formaldehyde, Pharos Reviews, Formaldehyde-Free Insulations www.pharosproject.net/blog/show/63/formaldehyde-free-insulation [viewed 3.2.16]

80 Lent, T. (2010), op cit.

81 Polyurethane Foam Association www.pfa.org/faq.html [viewed 3.2.16]

82 Styrene, potential for human exposure www.atsdr.cdc.gov/toxprofiles/tp53-c6.pdf [viewed 3.2.16]

3 Emissions from materials

Why do we need to use hazardous chemicals?

> Chemicals have enriched the lives of people throughout the world and improved their productivity and enjoyment of life. However, many of these same chemicals can pose significant risks to human health and too often disrupt and compromise the careful balances of natural ecosystems ... The problem with synthetic chemicals is similar to the problem with many other technologies – we rush to develop and enjoy the fruits of novel technologies, long before we fully understand their consequences and create the systems needed to address their costs.[1]

The best way to ensure good indoor air quality is to minimise the use of hazardous chemicals in buildings, however this is not easy when most professionals in the construction industry have come to accept synthetic chemistry as the solution to many building problems. While traditional materials such as timber and bricks continue to be used, chemicals become more and more prevalent in buildings and thus affect the indoor environment. Geiser and the Green Chemistry movement, by contrast, argue that while chemicals have been useful, we need to be much more careful about what we use and do everything we can to avoid or reduce the use of hazardous chemicals.

When IAQ tests are carried out in buildings, the principle aim is to identify hazardous chemicals that may affect health. However such is the wide range of possible chemicals used in building materials that most IAQ testing has to sample a limited range. The chemicals may come from outside the building, from vehicle engine pollution, or central heating boilers, but the primary source is the materials used to construct the building. This chapter explores the issue of why we should be more careful about the chemicals that lie within construction materials.

Companies that sell chemical-based materials often claim that emissions levels are too low to be of concern but this is not born out by the evidence presented in this book. However architects and builders have been lulled into a false sense of security, as they know little about where the products that they use come from, or how they are produced or the complex formulations and combinations of chemicals that go into products. The interest for many architects is for the building being produced on time within budget and the materials arriving off the back of a truck, usually wrapped in synthetic chemical plastic (which gets thrown away into landfill). Exploring the origins and the chemistry behind these materials is of little concern unless there is a specific health and safety risk to building workers on site, and often even this is ignored. However for those who wish to know more about how chemicals are produced and the damage they are doing, the following may be helpful.

It is important to consider how society has become increasingly dependent on hazardous chemicals and how these are crucial to understanding IAQ problems. It is also necessary to question why it is necessary to use hazardous chemicals when it is possible to create buildings from natural materials that are largely free of hazardous chemicals and that use very little energy to create them. By minimising the use of hazardous chemicals in buildings, IAQ problems can be significantly reduced. Alternative products that contain minimal chemical inputs, such as those certified by Natureplus, can meet the needs of most buildings. 'Natureplus … evaluates the compatibility of construction products, especially in the usage phase, according to strict standards in order to actively promote those materials *which pose no risk to health* and are, in addition, conducive to a healthy room climate'.[2]

Details of Natureplus certification

The crucial aspect of Natureplus certification, relevant to this book, is the list of prohibited substances. By excluding hazardous and toxic materials from construction products that are certified, the client, specifier or occupants can be assured that emissions have been reduced to the absolute minimum.

The general list of Natureplus prohibited substances covers substances that are classified as carcinogenic, causing mutations or toxic to reproduction. Additional individual substances have been specified by Natureplus as non-desirable due to their environmental and health dangers and which one would not expect to find in a certified product. Any exceptions to the exclusion of substances contained within the general list of prohibited substances are only possible on the basis of a comprehensive scientific justification, which must be presented in a scientific evaluation report, which has been commissioned by Natureplus. The general list of prohibited substances is shown in Figure 3.1.

CLP-Regulations: Carcinogenic Cat. 1A and 1B. Mutagenic Cat. 1A and 1B, Toxic to reproduction Cat. 1A and 1B
Substances as per DSD 67/548/EEC C1 and C2, M1 and M2, R1 and R2 and as per national law (e.g. TRGS 905)
Substances as per MAK-lists III1 and III2 (German occupational exposure limits)
Substances in IARC groups 1 and 2a (International Agency for Research on Cancer)
Substances requiring official approval as per Appendix XIV of the REACH regulations. In addition, the general list of prohibited substances includes the following named substances and compounds, as long as they have not already been included in the aforementioned lists: Persistent Organic Pollutants: Aldrin, Dieldrin, DDT, Endrin, Heptachlor, Chlordan, HCB, Mirex, Toxaphen, PCB, Dioxine und Furane
Arsenic and arsenic compounds
Lead and lead compounds
Cadmium and cadmium compounds
Organotin compounds
Antimony trioxide
Hydro-fluorocarbons (HFC)
Organic halogen phosphates

Figure 3.1 Natureplus prohibited substances
Source: Natureplus[3]

Further information about substances and materials that are hazardous is included in later chapters, together with materials and products that are approved by Natureplus that can be used as alternatives to chemical-based materials.

Chemical products – life cycle analysis and the recycling myth

Architects and specifiers are largely preoccupied with the cost and performance of materials and possibly their environmental impact. Sometimes distance travelled and embodied energy is considered through LCA and this includes scrutinising manufacturing processes and sources of materials. However not all LCAs are holistic and many tend to focus purely on energy efficiency and CO_2 emissions, ignoring other factors.[4,5]

Recent adoption by some in the construction industry of the concept of 'cradle to cradle' (C2C) or the 'circular economy' is actively promoting the idea that most chemical and synthetic materials can be recycled and used again, even though there is very little evidence that they can be recycled in practice. For some C2C is the last word in environmental best practice but Braumgart and McDonough's work is highly flawed and misleading, failing to question the overuse of valuable fossil fuel resources and the use of hazardous chemicals, which cannot be reused and will inevitably end up as hazardous landfill. C2C provides greenwash approval of products that are bad for the environment.[6,7]

C2C certification is a financial success for the companies given the label by McDonough and Braumgart, but much of the certification is based on a forecast that materials will be recycled in the future without any real evidence that this will actually take place. So for instance Dow Chemicals Styrofoam product has C2C Silver certification based on the hope that slabs of this insulation, used in roofs and floors, can be recycled in the future. However such materials are likely to be contaminated with screeds and adhesives that make them almost impossible to salvage once demolition takes place. Expanded polystyrene (EPS) foam is a major contaminant in landfill but the EPS industry even claims that 'EPS is an ideal material for landfill'.[8] Also, Styrofoam® XPS Insulation is a building insulation product. '*The reusable boards* feature high compressive strength but are also lightweight and are designed to be easy to cut and install'.[9]

> In a surprising development, Dow Chemical's Styrofoam brand extruded polystyrene (XPS) insulation has been awarded Cradle-to-Cradle Silver certification by McDonough Braumgart Design Chemistry (MBDC). 'It is shocking that a product containing a persistent organic pollutant such as HBCD can be considered green,' said EBN Advisory Board member Arlene Blum, Ph.D. Designers expecting a C2C logo to provide assurance that a specific product is free of hazards may want to read the fine print.[10]

To give them credit, C2C also certify many natural, bio-based and environmentally friendly products that can be easily recycled or are biodegradable, but by grouping these into similar categories to chemical products it devalues the standard to a form of greenwash. As Blum says it is important to look beyond the labelling and read the fine print.

Metals like copper, steel, aluminium and zinc are commonly recycled, when these can easily be salvaged. Masonry rubble is also reused as aggregate and fill, but as soon as chemicals are involved, the circular economy becomes much more complex and hard to achieve. Even metal windows that have been powder coated with a petrochemical-based paint finish involve creating huge amounts of toxic waste as the coatings have to be stripped or burnt off before the metal can be recycled, though in recent years an industry has emerged to recycle the powder coatings themselves. The PVC industry claims to recycle PVC windows but in reality this encompasses only a tiny percentage according to the Scottish Civic Trust.

> In reality uPVC is rarely recycled and when it is the uPVC degrades so that a window frame can only contain a small percentage of recycled material. The process requires the addition of yet more chemical additives and stabilisers and is actually more expensive than producing new uPVC. 82% of uPVC goes to landfill 15% of uPVC is incinerated. Only 3% of uPVC is recycled.[11]

Synthetic carpets, PVC flooring materials, anything glued together usually ends up in an incinerator or landfill leading to further health and pollution problems as buildings are still demolished with little thought for reuse.

Carpet manufacturers claim to recycle carpet materials[12] but in practice much of it ends up in landfill. A disused brickworks in Thrunton, Northumberland was used to store 3,000 tonnes of old carpets but caught fire in 2013 and the fire burned for over a year. Residents were seriously concerned that toxic chemicals had been washed into the local watercourses. Some carpet products use hazardous backing materials and are heavily impregnated with flame-retardant chemicals. Carpet manufacturers have been leading exponents of the so-called circular economy, claiming that nearly all their materials are recycled, but the fire at the Northumbrian dump suggests that recycling claims may be exaggerated.[13]

Natural materials by contrast can either be salvaged or reused or biodegrade naturally into the soil, providing they have not been contaminated by chemicals and plastics.

Chemical production – accidents waiting to happen

It is necessary to be aware of the vastness of the multinational chemical industry and how central it is to the production of building materials. The chemical industry, largely dependent on quarried materials and oil/petrochemicals, makes a significant contribution to pollution and health problems throughout the world. While many chemicals are still manufactured or processed in western developed countries, an increasing amount is now made in developing countries where environmental regulations and controls are not as strong. PVC powder used to manufacture products in the UK may have come from South America for instance.

> An £8 million haul of cocaine smuggled into Britain was discovered by employees at a chemical factory in 2000. Workers at Wardle Storeys, in Lancashire, thought there was something suspect about a consignment of PVC resin imported from Cartagena, in Colombia. While the white powdered resin was being unpacked at the Earby site on Tuesday, several blocks of yellow cocaine were found.[14]

An accident database – 'FACTS' (Failure and Accidents Technical Information System) holds information on more than 25,700 industrial incidents involving hazardous materials or dangerous goods that have happened all over the world during the past 90 years,[15] but only when there are major disasters do the location and nature of chemical manufacture and storage come to public attention. For instance in August 2015 massive explosions occurred in Tianjin in China causing over a hundred deaths, massive injuries and widespread pollution.[16]

The explosions took place in a warehouse owned by Ruihai International Logistics, according to BBC news. The chemicals included calcium carbide, potassium nitrate and ammonium nitrate. It was suspected that the warehouse also contained 700 tonnes of sodium cyanide. Calcium carbide is used in the production of PVC plastics but it is also used in the production of acetylene. Acetylene gas is used as a ripening agent for fruit stored in warehouses but it is also used in building products. 'The most common use of acetylene is as a raw material for the production of various organic chemicals including 1,4-butanediol, which is widely used in the preparation of polyurethane and polyester plastics'.[17] 1,4-butanediol, is considered a hazardous substance and has been investigated by the WHO, though there is very little data on toxicity in humans.[18]

Another explosion in 2015 occurred in Huantai in Shandong, destroying the Runxing factory. The factory produced adiponitrile, a colourless liquid that releases poisonous gases when it reacts with fire, according to Xinhua. Adiponitrile is used to make synthetic fibres such as nylon. Synthetic fibres are used to reinforce some construction products.[19]

Polypropylene and other polymer materials, widely used in construction materials, have suffered from several fires and chemical emissions in Japan (1996), a plastic factory in Fairfield, California (2011) and a chemical complex in Houston, Texas (2000).

> One fatality has been confirmed as a result of an explosion and fire yesterday at Phillips' K-Resin styrene-butadiene copolymer (SBC) plant, located at the company's Houston Chemical Complex (HCC) near Pasadena, Texas in 2000 (Manufacturing polyethelene, polypropylene, etc.). At least 71 workers were injured in the third major explosion at the factory in the past 11 years.[20]

'In April 2013 an Ammonium Nitrate storage facility in the small town of West, Texas, caught fire and exploded with devastating force'.[21] In an explosion and fire at a Dutch Shell chemical plant in 2014 it was said there appeared to have been a leak of benzene, though it was not known whether it was methylbenzene or ethylbenzene. The factory makes oil-based chemicals for use in products that range from car components to insulation materials.[22]

> Two people have been injured after an explosion caused a major fire at a Royal Dutch Shell chemical plant in the Netherlands. Flames and smoke were seen billowing into the sky at the site in the town of Moerdijk, near the port city of Rotterdam. The plant produces styrene monomer and propylene oxides which are raw materials used to make plastics.[23]

Chemical accidents also happen in the UK. The most infamous was at Flixborough in 1974 when 28 workers were killed in an explosion that was caused by leaking cyclohexane, a chemical used in the production of synthetic fibres, solvents and paint removers. Cyanide is a highly poisonous substance used in a variety of industrial processes including plastics manufacture. Cyanide is released when some synthetic foam products burn.

> Firefighters using breathing apparatus and specialist gas-tight chemical suits have tackled a cyanide leak in north Devon in December 2010. Devon and Somerset Fire and Rescue Service said about 200 litres of potassium cyanide leaked into an overflow tank at Tyco Electronics at East The Water, near Bideford.[24]

The most infamous chemical explosion was in Bhopal, India, in December 1984, leading to up to 20,000 deaths over subsequent years. Methyl Isocyanate was the chemical released.[25]

> Isocyanates are a family of highly reactive, low molecular weight chemicals. They are widely used in the manufacture of flexible and rigid foams, fibers, coatings such as paints and varnishes, and elastomers, and are increasingly used in building insulation materials. Spray-on polyurethane products containing isocyanates have been developed for a wide range of retail, commercial, and industrial uses to protect cement, wood, fiberglass, steel and aluminum, including insulation, protective coatings for truck beds, trailers, boats, foundations, and decks.
>
> Isocyanates are powerful irritants to the mucous membranes of the eyes and gastrointestinal and respiratory tracts. Direct skin contact can also cause marked inflammation. Isocyanates can also sensitize workers, making them subject to severe asthma attacks if they are exposed again. There is evidence that both respiratory and dermal exposures can lead to sensitization. Death from severe asthma in some sensitized subjects has been reported. Workers potentially exposed to isocyanates who experience persistent or recurring eye irritation, nasal congestion, dry or sore throat, cold-like symptoms, cough, shortness of breath, wheezing, or chest tightness should see a physician knowledgeable in work-related health problems.[26]

'Methyl isocyanate is extremely toxic to humans from acute (short-term) exposure'.[27] In 1999, Dow chemicals, the manufacturer of Styrofoam insulation, bought Union Carbide, the company that had been responsible for the Bhopal Disaster. The BBC, in an investigative programme, raised questions as to why Dow did not appear to be honouring agreements to compensate the victims of the Bhopal disaster but Dow argued that they and Union Carbide were separate companies. This has been questioned in court.[28] Because of the Bhopal legacy, controversy raged around Dow sponsorship of the 2012 London Olympic and a number of architecture practices now refuse to specify Dow products.[29]

Monitoring hazardous chemicals and the EU 'REACH' Directive

The International Chemical Secretariat has developed the wonderfully named SIN (Substitute It Now) List. The SIN List is a free to download excel database of hazardous chemicals based on the EU REACH (Registration, Evaluation, Authorisation and Restriction of Chemicals) Directive. Some say that the REACH Directive is a flawed tool in that many hazardous chemicals are not yet included and the commercial chemical industry has been highly active in protecting its interests by limiting the impact of REACH, as discussed further in Chapter 12.

> The SIN (Substitute It Now!) List is a concrete tool to speed up the transition to a world free from hazardous chemicals. The chemicals on the SIN List have been identified by ChemSec as Substances of Very High Concern based on the criteria established by the EU chemicals regulation REACH.
>
> The aim of the SIN List is to spark innovation towards products without hazardous chemicals by speeding up legislative processes and giving guidance to companies and other stakeholders on which chemicals to start substituting.
>
> There is an urgent need to reduce the use of harmful chemicals in society. Let SIN inspire you and be a concrete tool in your work towards a world with less hazardous chemicals. The SIN List is publically available in our free-of-charge online database.[30]

The REACH Directive of the EU came into force in 2006/2007, all chemicals have to be registered by 2018 and it is meant to control hazardous substances. The directive has led to changes in the format and content of safety data sheets that companies are required to publish. In the past, the old health and safety data sheets were a useful source of information on the chemicals and contents of products but since the REACH Directive this is no longer the case. Rather than increasing protection of the public from hazards and toxic chemicals, REACH has made it possible for manufacturers to hide this information, and guidance on what must be declared is highly confusing.

> Five years after its adoption, the European Commission is preparing to review the controversial REACH regulation, which for the first time required chemical manufacturers to justify that their products are safe for consumers. It's a potential 'can of worms,' according to EU officials.
>
> From the moment it was tabled until its eventual adoption in 2006, the REACH regulation gave rise to one of the most epic lobbying battles in the EU's history, pitting green campaigners against the powerful chemicals industry.[31]

There is no doubt that chemical companies have been forced to improve their record on pollution and emissions due to stricter environmental controls. For example the BASF chemical works in Cramlington has installed a regenerative thermal oxidiser to comply with Environment Agency regulations by removing more than 99 per cent of benzene contained in exhaust gases discharged from the processing of active pharmaceutical ingredients.[32]

A report commenting on chemical industry reduction of emissions of VOCs by 9,000 tonnes in 1999 referred to various companies, including Associated Octel Co. Ltd., which reduced VOCs from its Elllesmere Port plant by 4,553 tonnes, 70 per cent down on 1998 levels. BP Chemicals Ltd in Hull, Dupont in Cleveland and Union Carbide in Middlesborough achieved reductions of 65 per cent, 99 per cent and 55 per cent, respectively, together equivalent to a further 4,097 tonne reduction.[33] However reporting of such emissions issues seems to be much harder to find in recent years. The UK Environment Agency regulates emissions of VOCs and other chemicals such as hydrocarbons, plastic materials and di-isocyanates and issues permits, but data on pollution emissions is not so easy to find, compared for instance with US EPA reporting.[34]

BASF, one of the largest chemical companies in the world, is reported to have a history of accidental releases of hazardous chemicals in the USA. One occurred in 2013 when the Williams Olefin factory in Geismar, Louisiana killed two workers.[35] Olefins include ethylene, propylene and butadiene, all key chemicals in the manufacture of plastic and synthetic products.

Louisiana and the Mississippi River region are among the main areas for chemicals manufacture in the USA. There has been a long history of industrial disputes, worker lockouts and protests by local residents in the Mississippi River area due to major releases of phosgene, toluene and other toxic gases.

> Locked-out workers, using special air-monitoring devices, began random tests of air quality in areas around BASF and other chemical plants. A study documented that over a one-year period, 196 million pounds of chemical pollutants had been released by the 18 Geismar-area chemical plants. A joint study by OCAW and the Sierra Club further documented that 76 million pounds of chemicals had been dumped into the Mississippi River by 15 area plants in one year.[36]

The Mississippi River area from Baton Rouge to New Orleans is a favoured location for nearly 100 chemical plants and is known as 'chemical alley' or 'cancer alley' due to the high incidence of cancer in the area near the chemical plants. Dow Chemicals led the expansion of chemical factories along the Mississippi River in the 1950s and 1960s, partly due to strong trade union organisation moving them away from Texas and the Gulf Coast. The Mississippi River was populated by poor black people who were less well organised and old cotton plantations were quickly turned into one of the biggest concentrations of petrochemical plants in the world. One of the main industries was the production of chlorine and the area became a centre for PVC production. BASF became one of the largest companies based in the area.

> Cancer, birth defects and other conditions were documented among populations living near chemical plants where vinyl feed stock was being leaked and spewed into the water, air and soil. The crisis that loomed for the chemical industry was this: Is such a seemingly benign product as plastic could prove so dangerous and far-reaching, then what of the thousands of new chemicals that were introduced every year? By the early 1980s Louisiana displaced New Jersey and its chemical industry as the nation's most polluted state.[37]

BASF Aktiengesellschaft paid a US$225 million criminal fine in 1999. In Dallas, the Department of Justice charged the company with conspiring to fix, raise and maintain prices, and allocate the sales volumes of vitamins sold by them and other unnamed companies in the US and elsewhere.[38] While this issue is related to vitamins rather than building materials, it gives an indication of the corporate approach of the company. On another issue, oil company Shell and BASF paid $382 million into a compensation fund for over 1,000 workers following a ruling by a Brazilian judge. Workers claimed that they were contaminated by a chemical plant that operated from 1977 until it was closed in 2002. Shell originally owned it but sold the operation to American Cyanamid in 1995, which was bought by BASF in 2000.

> In its 2011 annual report, BASF acknowledged the site was 'significantly contaminated by the production of crop protection products.' But it denies responsibility, claiming that the site was contaminated before it bought the plant. Last year, BASF filed a lawsuit in Brazil asking a court to hold Shell fully responsible for any damages.[39]

The chemical industry has been very active in refuting claims about cancer and other health problems resulting from industrial pollution. For instance the Louisiana Chemical Association recruited an epidemiologist named Otto Wong, who claimed that the massive amount of data that had been collected about cancer in the Louisiana area was not due to emissions but the lifestyle of residents, implying that people there ate less vegetables and smoked and drank too much! However in 1991 Wong published an article drawing attention to significant mortality excesses in cancers, much to the fury of the chemical industry that had funded his research! He later retracted some of his findings.[40] However, whatever the academic and scientific debates, America's 'chemical heartland' has, according to Bullard, become a 'sump for the rest of the nation's toxic waste'.[41]

The influence of chemical companies on energy efficiency research

Leading petrochemical companies are not only highly active in lobbying the EU and other regulatory bodies, to ensure that there are not too many restrictions on the use of hazardous chemicals, but they also ensure that they have a significant influence on scientific research.[42] EU research programmes encourage industrial partners to be part of university-led research but industrial companies may be considered to have too much influence in such research, which fails to be sufficiently independent.[43]

In the UK a significant amount of scientific and technological research is funded by Innovate UK, previously known as the Technology Strategy Board. The TSB/Innovate has primarily been concerned to ensure that university research is largely industry led. Despite this, it might be expected that the board of Innovate UK would reasonably represent the wide range of industry and academic bodies that are funded by it, but this is far from the case. Instead the TSB's top people are mainly drawn from large multinational chemical and construction sector companies. Nearly everyone on the TSB/Innovate board is linked in some way to petrochemicals, cement, pharmaceuticals and polymers and similar. The other main academic research funding body, the Engineering and Physical Sciences Research Council (EPSRC), also includes members with interest in oil, chemical pharmaceutical companies and Rio Tinto, though it contains more academics.[44,45]

The manufacture of insulation and many other petrochemical and synthetic construction materials is dominated by a small number of large multinational construction companies. Not only do they play a big part in lobbying the European Commission through the European Chemical Industry Council (the largest) and organisations such as insulation manufacturers' body EURIMA (European Insulation Manufacturers Association) but they also play a big role in European-funded research and policy.[46] A scan of European-funded research projects on construction and energy efficiency will reveal that the large chemical companies play a significant role. For example, BASF is involved as a strategic partner of an EU-funded 'Energy Savings Topic Group' called EURHONET.[47] BASF also has an energy consultancy LUWOGE that prepares Passiv Haus certification for a project called Building Together (BuildTog). EU-funded BuildTog is a European Passiv Haus project initiated by four partners in EURHONET.[48] 'The objective of BuildTog is to establish the ground work for all European countries … to construct a generation of high-energy performance houses that are economically feasible and also architecturally ambitious'. BuildTog projects use only petrochemical-based insulations, as far as could be determined, and there was no evidence of any IAQ monitoring included in the research they have carried out.

Passiv Haus and the big chemical companies

Passiv Haus (PH) is the certification system for low-energy buildings that has become popular with architects and their clients and is closely connected with synthetic petrochemical product manufacturers. Some public bodies have tried to enforce PH as a standard for new construction through planning or building regulations. This became a controversial issue in Ireland with the Environment Minister Alan Kelly attempting to prevent some local authorities making PH compulsory for new build.[49] While it is technically feasible to build PH-certified buildings using natural non-hazardous materials, there are growing concerns about air quality problems with PH (discussed in Chapter 8).

Examples of manufacturing companies and product distributors that are patrons or members of PH[50] include:

- Kingspan
- St Gobain
- BASF

- Profine (PVC)
- Proctor Group
- Airtight tapes/SIGA
- Alsecco
- Ecoslab (Jablite)
- Icynene spray foam
- Isoquick
- REHAU PVC and Polymers

Government support for chemical materials in UK and Ireland

Support has been given by governments to support the manufacture and usage of chemical construction products, particularly insulation. The UK Green Investment Bank has invested £50 million in retrofitting UK industrial facilities owned and operated by Kingspan.[51] Local Authority Building Control and Energy Saving Trust advice to architects and householders tends to favour synthetic insulation materials while ignoring natural and renewable materials.[52]

While the Sustainable Energy Authority of Ireland does at least mention natural insulation materials, such as sheep's wool, in its guidance, it also exhibits a tendency to promote chemical products. For instance an energy upgrade of 2,100 homes in the Louth Energy Pilot project in northeast Ireland completed in 2014 (the Climote project) used Kingspan insulation.[53]

The northern counties of the Republic of Ireland, close to the border with Northern Ireland, have a surprisingly large cluster of factories and businesses making and distributing chemical-based insulation products, though its not clear why this area has attracted so many companies as there is no local source of raw materials. This is an area with a relatively low density of population and it is possible that there have been generous grants and incentives to bring employment to the area.

While many argue that society gets enormous benefits from chemicals, we are rapidly becoming dependent on materials synthethised from petrochemical resources even though these are regarded as a valuable and finite resource. As Carson and Mumford say in their 600-page book on hazardous chemicals, 'The hazards of chemicals stem from their inherent flammable, explosive, toxic, carcinogenic, corrosive, radioactive or chemical-reactive properties. Society must strike a balance between the benefits and risks of chemicals'.[54]

In buildings we can minimise the use of synthetic oil derived materials relatively easily and it is mainly those promoting technology-based solutions to energy efficiency that force the industry to use so many chemical-based products. This exposes us all to toxic hazards in the general environment as well as in our buildings.

Notes

1 Geiser, K. (2015) *Chemicals Without Harm: Policies for a Sustainable World*, IT Press

2 Natureplus, Basic Criteria www.natureplus.org/fileadmin/user_upload/pdf/cert-criterias/RL00 Basiskriterien_en.pdf [viewed 12.2.16]

3 Natureplus, Basic Criteria, op cit.

4 Bribián, I.Z., Capilla, A.V. and Usón, A.A. (2011) Life cycle assessment of building materials: Comparative analysis of energy and environmental impacts and evaluation of the eco-efficiency improvement potential, *Building and Environment* 46 (5), 1133–1140

5 Thiel, C.L., Campion, N., Landis, A.E., Jones, A.K., Schaefer, L.A. and Bilec, M.M. (2013) A materials life cycle assessment of a net-zero energy building, *Energies* 6, 1125–1141; doi:10. 3390/en6021125

6 Braungart, M. and McDonough, W. (2002) *Cradle to Cradle: Remaking the Way We Make Things*, North Point Press

7 Circular economy, Ellen Macarthur Foundation www.ellenmacarthurfoundation.org/circular-economy [viewed 12.2.16]

8 www.eps.co.uk/sustainability/recycling_practicalities.html [viewed 24.3.16]

9 Cradle to Cradle Certified Products Registry www.c2ccertified.org/products/registry/search&keywords=insulation [viewed 5.2.16]

10 Dow Styrofoam Cradle to Cradle Certified, Despite Chemistry (2010) *Environmental Buildings News* 19 (2) www2.buildinggreen.com/article/dow-styrofoam-cradle-cradle-certified-despite-chemistry-0 [viewed 12.2.16]

11 Scottish Civic Trust A How to Guide (2008) Windows: Repair and Replacement www.scottishcivictrust.org.uk/media/160859/windows_16.4.14.pdf [viewed 12.2.16]

12 Carpet Recycling UK www.carpetrecyclinguk.com/ [viewed 5.2.16]

13 Frustrations of residents living near 'burning tip', *Northumberland Gazette* www.northumberlandgazette.co.uk/news/local-news/frustrations-of-residents-living-near-burning-tip-1-6257849 [viewed 5.2.16]

14 Factory workers find pounds 8m haul of cocaine, The Free Library www.thefreelibrary.com/Factory+workers+find+pounds+8m+haul+of+cocaine.-a060818068 [viewed 12.2.16]

15 Failure and Accidents Technical information System www.factsonline.nl/ [viewed 12.2.16]

16 China explosions: What we know about what happened in Tianjin, BBC News www.bbc.co.uk/news/world-asia-china-33844084 [viewed 5.2.16]

17 Acetylene, How products are made www.madehow.com/Volume-4/Acetylene.html [viewed 5.2.16]

18 1,4-Butanediol (1,4-BD) Pre-Review Report, WHO www.who.int/medicines/areas/quality_safety/5.51-4Butanediolpre-review.pdf [viewed 5.2.16]

19 China rocked by second deadly chemical plant blast, CNBC www.cnbc.com/2015/08/23/china-rocked-by-second-deadly-chemical-plant-blast.html [viewed 5.2.16]

20 Chemical Plant Explosion Kills 1, Injures 71, *EHS Today* 71 http://ehstoday.com/news/ehs_imp_33143 [viewed 5.2.16]

21 Pearce, N. (2015) Safer storage, *NFPA Journal* www.nfpa.org/newsandpublications/nfpa-journal/2015/may-june-2015/features/nfpa-400 [viewed 5.2.16]

22 Explosion and fire at Dutch Shell chemical plant, BBC News www.bbc.co.uk/news/world-europe-27691421 [viewed 5.2.16]; Toxic chemical cloud hits Russian city, SFGATE www.sfgate.com/world/article/Toxic-chemical-cloud-hits-Russian-city-2311749.php [viewed 5.2.16]

23 Fire breaks out at Shell plant in Netherlands, Sky News http://news.sky.com/story/1274905/fire-breaks-out-at-shell-plant-in-netherlands [viewed 5.2.16]

24 Fire-fighters tackle Bideford cyanide gas leak, BBC News www.bbc.co.uk/news/uk-england-devon-11974796 [viewed 5.2.16]

25 International campaign for justice in Bhopal www.bhopal.net/ [viewed 5.2.16]

26 Isocyanates, Centre for Disease Control and Prevention www.cdc.gov/niosh/topics/isocyanates/ [viewed 5.2.16]

27 Methyl isocyanate, EPA www.epa.gov/airtoxics/hlthef/methylis.html [viewed 5.2.16]

28 New evidence favours Bhopal gas victims battling Dow Chemical, Down to Earth www.downtoearth.org.in/news/new-evidence-favours-bhopal-gas-victims-battling-dow-chemical-35894 [viewed 5.2.16]

29 Response: Union Carbide and Dow Chemical, BBC News http://news.bbc.co.uk/1/hi/programmes/bhopal/4023447.stm [viewed 5.2.16]

30 Chemsec, SIN List http://chemsec.org/what-we-do/sin-list [viewed 5.2.16]

31 EU prepares to re-open REACH 'can of worms', EurActive www.euractiv.com/climate-environment/eu-prepares-open-reach-worms-news-507129 [viewed 5.2.16]

32 VOC abatement plant ensure compliance at chemical works, Edie.net www.edie.net/library/VOC-abatement-plant-ensure-compliance-at-chemical-works/491 [viewed 5.2.16]

33 Pollution failures mar improving record, Edie.net www.edie.net/library/Pollution-failures-mar-improving-record/1529 [viewed 5.2.16]
34 www.gov.uk/guidance/environmental-permits-licences-and-exemptions-for-chemical-manufacturing [viewed 5.2.16]
35 Williams Olefins Plant explosion and fire, US Chemical Safety Board www.csb.gov/williams-olefins-plant-explosion-and-fire-/ [viewed 12.2.16]
36 Leonard, R. and Nauth, Z. (1990) Beating BASF: OCAW busts union-buster, *Labor Research* 1 (16), Article 8
37 Markowitz, G. and Rosner, D. (2013) *Deceit and Denial: The Deadly Politics of Industrial Pollution, Ol Man River or Cancer Alley?*, Chapter 8, University of California Press
38 Top 100 Corporate Criminals of the Decade, Information Clearing House www.informationclearinghouse.info/article3676.htm [viewed 5.2.16]
39 Shell, BASF convicted for pollution, Dear Kitty, Some Blog https://dearkitty1.wordpress.com/2012/07/03/shell-basf-convicted-for-pollution/ [viewed 5.2.16]
40 Markowitz, G. and Rosner, D. (2013) op cit.
41 Bullard, R. (1990) *Dumping in Dixie: Race class and environmental quality*, Westview Press
42 Wenger, C., Secher, T., Mohammed, H., Alizadeh, A. and Khosravi, P. (2015) *Disarming REACH, An Insight to Lobbying in the European Union* www.academia.edu/8515935/Disarming_REACH_-_An_Insight_to_Lobbying_in_the_European_Union [viewed 12.2.16]
43 EU research funding: for who's benefit?, CEO Corporate Europe http://corporateeurope.org/sites/default/files/publications/research_report_-_final.pdf [viewed 12.2.16]
44 www.gov.uk/government/publications/innovate-uk-board-members-interests [viewed 5.2.16]
45 www.epsrc.ac.uk/about/governance/council/ [viewed 5.2.16]
46 www.eurima.org/ [viewed 5.2.16]
47 (EURHNET) www.linkedin.com/company/eurhonet-european-housing-network [viewed 5.2.16]
48 What a 'Buildtog' is www.buildtog.eu/fileadmin/downloads/150306-What-a-BuildTog-is.pdf?1425628688 [viewed 5.2.16]
49 Ministers warn Dublin councils against new housing standards, *Irish Times*, August 24, 2015
50 Passiv Haus Trust www.passivhaustrust.org.uk [viewed 5.2.16]
51 Sustainable Development Capital agrees first energy efficiency investment for UK government's Green Investment Bank, SDCL www.sdcl-ib.com/sustainable-development-capital-agrees-first-energy-efficiency-investment-for-uk-governments-green-investment-bank/ [viewed 5.2.16]
52 Energy Saving Trust www.energysavingtrust.org.uk/domestic/products?field_product_category%5B0%5D=4846 [viewed 12.2.16]
53 Minister Alex White reviews completion of the Louth Energy Pilot, Climate News Project www.climote.ie/2015/07/14/minister-alex-white-reviews-completion-of-the-louth-energy-pilot-project/ [viewed 5.2.16]
54 Carson, P. and Mumford, C. (2002) *Hazardous Chemicals Handbook*, Second Edition, Butterworth Heinemann

4 Cancer, carcinogens and building materials

Can building materials contribute to an increased risk of cancer? A significant number of modern chemical-based construction materials contain chemicals that are known as carcinogenic and thus can possibly cause cancer. It is claimed by Cancer Research UK that half of us will get cancer at some point in our lives and this risk has increased in recent years. The increase in cancer is usually attributed to smoking (even though this has significantly reduced in recent years) ageing, obesity and diet.[1]

The increase in cancer is not commonly attributed to environmental problems, IAQ and hazardous emissions, but this is not because there is no significant risk, but because the issue has been largely ignored by medical researchers and cancer charities. While smoking, obesity and less healthy lifestyles may have contributed significantly to the huge increase in cancer, this cannot possibly account entirely for its exponential rise. Indeed lifestyles have been changing for the better, alcohol consumption in the UK has fallen and many more people have become exercise and health conscious. However, pollution and IAQ have got worse. Thus some experts and authorities have recently been forced to draw attention to environmental causes, suggesting that these play a much more significant role in cancer occurrence than had been previously recognised.

This chapter is about the environmental causes of cancer and emissions from carcinogenic materials used in buildings. It will be relatively easy to establish a link between cancer and hazardous materials but less easy to demonstrate how prevalent these are in buildings. Manufacturers will certainly argue that their materials, which they are reluctant to admit are hazardous, are not present in sufficient concentrations in buildings to provide a health risk. Common sense would suggest that hazardous materials when locked up in building fabric and isolated from indoor air should not be a problem and yet when IAQ tests are carried out, formaldehyde and other carcinogenic materials are readily detected and often in surprisingly high concentrations. Indeed some mainstream companies are now advertising materials that they claim can remove VOCs and formaldehyde from the indoor environment. They have spotted a market opportunity, as have other companies that promote low-formaldehyde alternative insulations and composite products.

Recognising the environmental causes of cancer

The bodies concerned with cancer in the UK underplay the problems of hazardous materials, for instance, Cancer Research UK, and while recognising environmental and occupational causes of cancer, they do not refer to indoor air pollution as a significant factor, other than radon and smoking, on their websites and guidance documents. External air pollution is given greater weight.[2]

Cancer Research UK does recognise cancer risks to workers from construction and painting but provides no details of the risks on their website.

There is little doubt that there has been a significant worldwide growth of cancer among all populations in recent years, but this is largely explained in terms of bad diet and an increase in ageing populations.

> Global cancer rates could increase by 50% to 15 million by 2020 … The predicted sharp increase in new cases – from 10 million new cases globally in 2000, to 15 million in 2020 – will mainly be due to steadily ageing populations in both developed and developing countries and also to current trends in smoking prevalence and the growing adoption of unhealthy lifestyles.[3]

However, WHO gives some recognition to IAQ problems as a cause of cancer and does recognise it as a problem. 'Environmental pollution of air, water and soil with carcinogenic chemicals accounts for 1–4% of all cancers (IARC/WHO, 2003). Exposure to carcinogenic chemicals in the environment can occur through drinking water or *pollution of indoor* and ambient air'.WHO refers to significant problems of cancer as a result of cooking fuels and combustion methods, mostly in developing countries, but in some reports its general guidance, implies that IAQ problems are not a significant factor in cancer rates.[4] WHO guidelines on IAQ however make specific reference to minimising risks from carcinogenic materials such as formaldehyde, naphthalene and PAHs (polycyclic aromatic hydrocarbons). Risks of cancer from these and other hazardous materials are clear in the details of WHO research but explicitly linking cancer with hazardous IAQ problems is avoided, though occupational exposure is cited. Benzene (a cause of leukemia) in indoor air as a cause of cancer is stated clearly however. 'Benzene is a genotoxic carcinogen in humans and no safe level of exposure can be recommended. The risk of toxicity from inhaled benzene would be the same whether the exposure were indoors or outdoors'.[5] And, 'The use of low-emitting building materials and products, and preventing exposures to environmental tobacco smoke and other combustion emissions, will minimize exposure-related risk. In addition, ventilation can reduce indoor exposure to formaldehyde'.[6]

The guidelines from WHO can be confusing as in some they refer to hazardous indoor emissions from carcinogenic materials, but in others refer to insufficient medical and epidemiological studies. This makes it easier for environmental cancer deniers to ignore the problem.

Cancer charities in the UK

Early in the process of compiling this book I contacted all the leading cancer charities in the UK (ten in all) and asked them what information they had about environmental causes of cancer and whether they had policies of excluding carcinogenic materials from their own buildings. Some of the charities did not reply, despite numerous reminders, but some were very open and willing to discuss the issue and I met with three of the leading organisations and visited some of their new buildings. Some cancer care and hospice charities have major building programmes and might be expected to have a greater concern about IAQ issues. This was not the case and they used excuses for ignoring the issue of carcinogenic materials, such as only complying with the building regulations or leaving it to the architects. Not a single cancer charity that I was able to contact had any policies concerning the environmental causes of cancer *and none had policies in place to exclude carcinogenic materials from their buildings*. When offered the opportunity for a seminar to discuss existing knowledge of these issues, none were willing to consider this. A similar offer to some architects working on cancer care buildings was also rejected.

When asked why they had such a head-in-the-sand approach to the issue, the reason given by some was that they would not consider the issue unless presented with solid scientific evidence that environmental and construction chemicals caused cancer, however not one of them was willing to provide an opportunity for this evidence to be presented.

Environmental causes of cancer or 'bad luck'

A controversial study published in *Science* in January 2015 by Tomasetti and Vogelstein, claimed that 'Only a third of the variation in cancer risk among tissues is attributable to environmental factors or inherited predispositions. The majority is due to "bad luck," that is, random mutations arising during DNA replication in normal, noncancerous stem cells'.[7] Admitting that up to one third could be attributed to environmental factors was a major step forward from earlier literature. However the 'bad luck' hypothesis was challenged in a major review published in *Nature* and given a great deal of publicity.[8] The study carried out by Hannun and others at the Stony Brook Cancer Centre in New York, USA, reversed the conclusions of the Tomasetti report and claimed that *environmental causes may account for up to 90 per cent of cancers*. They 'found quantitative evidence proving that extrinsic risk factors, such as environmental exposures and behaviors weigh heavily on the development of a vast majority (approximately *70 to 90 percent)* of cancers'.[9]

> These results are supported by strong epidemiologic evidence; for example studies showing that immigrants moving from countries with lower cancer incidence to countries with higher rates of cancer incidence acquire the higher risk in their new country. The majority of cancers, such as colorectal, lung, bladder and thyroid cancers had large proportions of mutations likely caused by extrinsic factors.
>
> The team also analyzed the SEER (Surveillance, Epidemiologic and End Results Program) data, which showed that many cancers have been increasing in incidence and in mortality, suggesting that external factors contribute heavily to these cancers.[10]

This and other recent studies indicate that attitudes to the environmental causes of cancer are now changing. 'How many cancers are caused by the Environment? Some experts say a decades-old estimate that six percent of cancers are due to environmental and occupational exposures is outdated and far too low'.[11]

While the debate between proponents of the just 'bad luck' and environmental camps is likely to continue for some time, there seems enough evidence to suggest that the environmental causes of cancer should be taken seriously. Environmental causes can of course mean a wide range of things, including smoking, external air pollution and lifestyle factors, as well as IAQ, so it is necessary to examine the literature about how toxic chemicals in buildings might cause cancer.

Scientific evidence of how toxic chemicals in buildings causes cancer

Rager *et al.* give some indication as to how airborn toxic chemicals can cause cancer. Using formaldehyde as an example, they explain that it has been long known that formaldehyde can cause irritation and nasopharyngeal cancer but their study explains how it has an impact on 'signalling pathways' related to inflammation, cancer and endocrine disruption: 'this research suggests a novel epigenetic mechanism by which formaldehyde may induce disease ... Our study provides evidence of a potential mechanism that may underlie the cellular effects induced by formaldehyde, namely, the modification of miRNA expression'.[12]

Medical research such as that carried out by Rager and her team at the University of North Carolina, Chapel Hill, shows how chemicals can affect genetic and DNA processes, and how toxic chemicals cause cancer, thus providing a methodology to examine links and causation. Scientific work such as this raises considerable concern about the risks from carcinogenic materials but much more work needs to be done, as it is hard to separate out particular chemicals in medical research when so many people are subject to breathing in a wide range of toxins. There has been insufficient work for instance to compare the risks of cancer from occupational (workplace) exposure with domestic exposure. Exposure to lower level emissions, but over a longer period of time, may be as big a risk as short-term exposure to higher doses in the workplace, but insufficient research has been done on this. Also much published risk assessment is based on tests on rats and laboratory tests rather than people in real buildings.

The American Lung Association warns of the dangers of chemicals affecting IAQ but stops short of linking formaldehyde with cancer.

> A common chemical, (formaldehyde) found primarily in adhesive or bonding agents for many materials found in households and offices, including carpets, upholstery, particleboard, and plywood paneling. The release of formaldehyde into the air may cause health problems, such as coughing; eye, nose, and throat irritation; skin rashes, headaches, and dizziness.[13]

However the International Agency for Research on Cancer (IARC), a division of WHO, has made very clear links between formaldehyde and cancer and also spells out building materials as a source.

> Twenty-six scientists from 10 countries evaluated the available evidence on the carcinogenicity of formaldehyde, a widely used chemical. The working group, ... concluded that formaldehyde is carcinogenic to humans. Previous evaluations, based on the smaller number of studies available at that time, had concluded that formaldehyde was probably carcinogenic to humans, but new information from studies of persons exposed to formaldehyde has increased the overall weight of the evidence.
>
> Formaldehyde is produced worldwide on a large scale. It is used mainly in the production of resins that are used as adhesives and binders for wood products, pulp, paper, glasswool and rockwool. Formaldehyde is also used extensively in the production of plastics and coatings, in textile finishing and in the manufacture of industrial chemicals ... Common sources of exposure include vehicle emissions, particleboards and similar building materials, carpets, paints and varnishes, foods and cooking, tobacco smoke, and the use of formaldehyde as a disinfectant. Levels of formaldehyde in outdoor air are generally low but higher levels can be found in the indoor air of homes.[14]

In the USA there seems to be a greater level of awareness and concern about cancer risks when compared with the UK. Physicians for Social Responsibility (PSR) is a US organisation over 50 years old. It is linked to the International Society of Doctors for the

Environment. PSR list the following examples of environmental carcinogens found in food and buildings and point out that cancer accounts for one in four deaths and 1,500 deaths per day in the USA from over a hundred types of cancer:[15]

- Arsenic
- Asbestos
- Benzene
- Bisphenol A (BPA)
- Chromium hexavalent compounds
- Dioxins
- Formaldehyde
- Polybrominated diphenyl ethers (PBDEs)
- Polycyclic aromatic hydrocarbons (PAHs)
- Vinyl chloride[16]

PSR draws attention to the US President's Cancer Panel, which reported that

> the true burden of environmentally induced cancers has been grossly underestimated. Exposure to environmental carcinogens (chemicals or substances that can lead to the development of cancer) can occur in the workplace and in the home … It has long been known that exposure to high levels of certain chemicals, such as those in some occupational settings, can cause cancer. There is now growing scientific evidence that exposure to lower levels of chemicals in the general environment is contributing to society's cancer burden.
>
> Because something has been classified as a carcinogen does not mean that every instance of exposure to that substance will result in the development of cancer. By the same token, a listing of 'probable' or 'possible' carcinogenicity does not mean we have exhausted study on that substance. It means the substance is not yet sufficiently studied. Such substances may, with further study, turn out to be definitively carcinogenic.[17]

The President's Cancer Panel has stated that a ubiquitous chemical, BPA, is still found in many consumer products (and building materials) and remains unregulated in the US, despite the growing link between BPA and several diseases, including various cancers. It goes on to say that the public remains unaware of many common environmental carcinogens such as formaldehyde and benzene and points out that children are far more vulnerable to environmental toxins and radiation than adults. 'All levels of government, from federal to local, must work to protect every American from needless disease through rigorous regulation of environmental pollutants'.[18]

Given the strength of this warning, it is important to be aware of how many possible environmental pollutants exist and which are likely to be found in buildings. The most definitive list comes from the IARC, which provides a list of chemical agents with several hundred classified as 1, 2A and 2B carcinogens. Agents classified as carcinogenic by the IARC Monographs, volumes 1–113 show that in Group 1 'Carcinogenic to Humans', 117 are listed. In Group 2a 'Possibly Carcinogenic to Humans', 74 are listed and in Group 2b 'Possibly Carcinogenic to Humans', 287 are listed.[19] Many of the chemicals and substances listed by the IARC are used in building materials. While the evidence of cause and effect in terms of cancer is not yet firmly established, there is plenty of solid scientific research that confirms their carcinogenity.

An even more extensive list comes from the US National Toxicology Program 13th *Report on Carcinogens* in 2014. This 18 megabyte document provides profiles of 187 substances that are reasonably anticipated to be a human carcinogen or confirmed carcinogens.[20] This extensive database lists the use of the chemicals and it is immediately clear how complex the use of such chemicals is, as they are frequently combined with many other chemicals to create different products. For example, 51 chemicals from this list that are directly involved in the production of construction materials and finishes, have been extracted from this list and can be used as a check against information that can be obtained from manufacturer declarations and environmental certificates (see Appendix A).

Formaldehyde

Annually, 5 million tonnes of formaldehyde are manufactured and production of formaldehyde is a major industry in the US, creating thousands of jobs.[21]

Formaldehyde is used in a wide range of products including cosmetics, food preservatives and building products. It has been linked with leukaemia and cancer, particularly in occupational studies.[22]

Major indoor sources of formaldehyde include some paints, carpets, composite wood products, laminate flooring and insulation materials. When carrying out IAQ tests, formaldehyde levels are always surprisingly high, even in buildings that have been designed to be environmentally friendly and use low toxicity natural materials. Formaldehyde is also present in natural wood as well as composite materials and the risk of formaldehyde emissions is sometimes obscured by suggestions that natural wood contains formaldehyde. However scientific tests show that formaldehyde emissions from natural wood are extremely low and significantly lower than emissions from composite glued products. The highest emissions values for natural wood are from oak are between 2 and 9 parts per billion (ppb) at 30 degrees C. Douglas fir, spruce and pine are generally 3–5 ppb.[23] Thus emissions from natural wood range from 2 to 9 ppb, whereas 'safe' levels in the UK and Ireland for formaldehyde emissions are 1 part per *million* (ppm), which is a thousand times higher. IAQ tests often reveal formaldehyde emission levels well above 1 ppm (even as high as 100 ppm). Thus emissions from natural wood can be regarded as insignificant.

Composite wood products are seen as having environmental benefits because it is possible to use poorer quality wood, thus reducing pressure on the felling of high quality timber. A range of products from structural beams, a wide range of boards, flooring and many other products are made with wood chip, wood waste or thin strips of softwood. However most of these products (not all) require glues and these have traditionally been made from synthetic chemicals, which include or emit formaldehyde. Composite and laminate wooden flooring products can be of particular concern as glues may also be used to fix them down. Mechanical fixing of flooring should always be used if possible. Germany and California[24] have laid down stringent regulations for composite wooden flooring in order to limit formaldehyde emissions. The US EPA has also published guidelines [25] and legislation was introduced, in the USA, to control toxic chemicals in wood products [26,27]

Literature on formaldehyde emission levels can be very confusing because of the different forms of measurement used in different scientific papers and standards, such as $\mu g/m^3$ (micrograms per cubic metre) or mg/m^3 (milligrams per cubic metre) or ppm (parts per million) or ppb (parts per billion). Concentrations of gaseous pollutants are sometimes given in units of mass per volume, and at other times in ppb or ppm or in $\mu g/m^3$, which is micrograms of gaseous pollutant per cubic metre of ambient air. Frequently, safe levels are set in

100ng/L = 0.1μg/l
1μg/L = 1ppb
mg/m³ = 0.1μg/l
(1g/m³ = 1 mg/L)
0.1 μg/L = 100 ng/L = 0.1 mg/m³

Figure 4.1 Conversions of methods of measurement

Note: Conversion to parts per million is more complicated

Source: DMU[28]

terms of ppm but emission figures are often given in micrograms per cubic metre (mg/m³ or μg/m³)! IAQ readings often give results in ng/L (nanograms per litre).

Measurements of any emissions in air are also affected by temperature and different figures are given for emissions over certain timeframes, such as one hour or a day at different temperatures!

Because of the confusion caused by the use of different measurements, it is hard to interpret published figures when considering IAQ standards and emission test results. For instance it is hard to correlate emissions from specific materials with potential IAQ levels. IAQ levels will be affected by a cocktail of materials, such as formaldehyde emitted from several different materials in the same room. The only way to be sure of good IAQ is to use materials that are free of formaldehyde. However products labeled low or zero formaldehyde cannot necessarily be trusted as there is not yet a finally agreed standard or methodology for testing or labeling and so different manufacturers will use different methods. Legislation signed by President Obama in the USA in 2010 sets limits for formaldehyde emissions from wood products and materials in the USA.[29] Various other organisations have set standards for formaldehyde emissions in IAQ. Standards for formaldehyde emissions from composite wood products have also been established in Europe, though probably only enforced in Germany. Many new products have been developed with alternatives to formaldehyde glues as a result of these standards, but reliable information from manufacturers is hard to find.

> Since 2004: Emission classes E1 and E2 were established by European Standard EN 13986 for use in construction where formaldehyde-containing materials, particularly resins, have been added to the product as a part of the production process, the product shall be tested and classified into one of two classes: E1 and E2 0.05 ppm boards can be marked with an environmental label (Blue Angel) 0.03 ppm boards are obligatory for members of the German Association of Producers of Prefabricated Houses BDF (since 2003).[30]

Medium-density fibreboard (MDF)	17,000–55,000 μg/m³ per day
Hardwood Plywood	1,500– 34,000 μg/m³ per day
Particleboard (chipboard)	2,000–25,000 μg/m³ per day
Urea formaldehyde foam insulation	1,200–19,200 μg/m³ per day
Fibreglass products	400–470 μg/m³ per day

Figure 4.2 Range of formaldehyde emissions rates in μg/m³ per day

WHO	below 0.10 ppm
European E1 standard	below 0.10 ppm
Green Guard and State of California	0.05 ppm
BREEAM HEA 02	0.10 mg/m^3

Figure 4.3 Different formaldehyde emission standards

While it may appear that 0.03 to 0.05 ppm is a tiny amount that might not be noticed in a house, constant exposure over time to formaldehyde, even at these levels may represent a health risk and thus formaldehyde should be kept to a minimum as much as possible. As can be seen from Figure 4.2, products such as MDF and chipboard can emit very high levels of formaldehyde in a day. Tests on most modern buildings in the UK will commonly reveal levels over 0.05 ppm and can go over 0.1 ppm.

A study by McGill *et al.*, of eight recently built housing association houses constructed to BRE Code Levels 3 and 4 in Northern Ireland, found formaldehyde levels above what is generally accepted as safe (0.08 ppm) and in some houses the level was as high as 1.85 ppm. It was initially suspected that, as these levels appeared to be intermittent, they might be due to candles or incense but as the occupants kept diaries, it was clear that some of the highest formaldehyde levels occurred when the houses *were unoccupied.*

> For instance, in C4: No.3, occupants stated a plug-in incense/scented candle was utilised from approximately 6 pm to 1 am on day one, and from 9 am to 4 pm on day two. Cleaning products were utilised from approximately 2 pm to 4 pm on day two, and the heating was turned on from 10 to 12 pm on day one and 3 to 4 pm on day two. Time and duration of these activities do not correspond with the peaks of formaldehyde.
>
> In C3: No.1, occupants stated no-one was home on day two from approximately 1 to 4 pm, during which time the significant peak of formaldehyde occurred. This is considerably unusual, and suggests either the source of formaldehyde was not occupant related, or the use of the occupant diary did not allow for accurate recording of activities during the monitoring period. Results from the living room carbon dioxide measurements however demonstrate a steady decline in levels from 13:35 to 16:05, suggesting that the occupants had indeed left the home.[31]

The BREEAM HEA 02 limiting level of 0.1 ppm may be regarded as too high to ensure good IAQ, as other authorities have set a safe limit at half this. Test results showing lower levels of formaldehyde may lull occupants into a false sense of security. Adopting the precautionary principle should lead to the avoidance of formaldehyde.

Urea formaldehyde

Urea formaldehyde foam insulation (UFFI) was quite commonplace for many years throughout the world. The US National Cancer Institute warned of the serious cancer risks from UFFI in 2011 but assumed that the use of urea insulation was uncommon, as efforts had been made to ban UFFI in USA and Canada. Their main concern was about dangers to workers rather than in houses.[32]

UFFI has been widely used in the UK particularly in social housing, despite its high toxicity. Inspection of Northern Ireland Housing Executive properties in the Rathfriland area of County Down revealed that UFFI had been used to insulate cavity walls in the 1970s but the insulation had significantly deteriorated.[33] Due to these concerns about cavity wall insulation in Northern Ireland a study revealed that both fibreglass and UFFI needed to be replaced.

> 4.10.4 Foamed Insulants: Urea-Formaldehyde Foam Insulation (UFFI) The use of urea-formaldehyde foam insulation (UFFI) urea-formaldehyde, also known as urea-methanal, dates back as far as the 1930s. It is basically a foam, not unlike shaving cream that is easily injected into walls under pressure. It is made by using a pump set and hose with a mixing gun to mix the foaming agent, resin and compressed air. The fully expanded foam is pumped into cavities in need of insulation. It becomes firm within minutes but cures within a week. UFFI is generally spotted in homes built during the 1970s and on. Visually it looks like oozing liquid that has been hardened.[34]

UFFI has been used in many homes. When installed it has a significant odour but this can still be detected many years later. 'Residents of UFFI insulated homes frequently complain of headache, malaise, insomnia, anorexia, irritation of the upper respiratory tract and loss of libido'.[35] Despite the fact that serious health problems have been associated with UFFI for many years, its use is still permitted in some countries and even advocated by some: 'We believe that those who have urea formaldehyde foam insulation in their homes should enjoy their houses, and sleep well at night'.[36] The use of UFFI has been permitted in the UK building regulations providing 'there is a continuous barrier which will minimize as far as practicable the passage of fumes to the occupiable parts' and 'Reasonable precautions should be taken to prevent the permeation of fumes'.[37]

Part B of the Northern Ireland Building Regulations prohibits the use of urea formaldehyde except in cavity insulation.[38] However, according to a consultation document issued in the UK in 2012, urea formaldehyde was still being installed in 700 buildings a year.[39]

UFFI was banned in Canada and the USA in the 1980s by which time over 100,000 houses had been treated in Canada. There are still installers in Europe that offer and install UFFI and it is surprising that its use is still permitted. When renovation work is being done to existing buildings, UFFI may be found. It should be treated as a hazardous material and removed with care, with full safety protection worn, as there are emissions of ammonia as well as formaldehyde. The UK Health and Safety Executive issues a wide range of guidelines about the risks from formaldehyde and UFFI.[40]

Styrene

Compounds containing styrenes have been listed by the National Toxicology Programme as a possible carcinogen, but this remains a controversial judgement.[41] Styrene is not only emitted from polystyrene and a wide range of polymer products but is also emitted from wood and cigarette smoke.

Scientific studies remain inconclusive. Styrene is metabolised into styrene oxide, which is classified as carcinogenic as a result of animal tests.

Styrene butadiene is a common compound used in tyres and found in a number of synthetic flooring materials and is regarded as a serious external air pollutant.

> The limited evidence for cancer from styrene in humans is from occupational studies showing increased risks for lymphohematopoietic cancers, such as leukemia and lymphoma, and genetic damage in the white blood cells, or lymphocytes, of workers exposed to styrene. There is also some evidence for increased risk of cancer in the pancreas or esophagus among some styrene workers.[42]

Benzene

Benzene is a serious pollutant and carcinogen. It causes chromosomal aberrations leading to cancer. Leukaemia is particularly associated with benzene exposure in the workplace. Exposure to solvents and working in the oil and petrol industries is a major source. Principle sources of benzene emissions are from car exhausts, vehicle fuel, residue from vehicle tyres and their manufacture, and other industrial pollution. Benzene is a common constituent of external air pollution. Concentrations are high around petrol stations and are associated with traffic pollution. Benzene is emitted by heating and cooking fuels and garages attached to houses are a particular risk. Benzene is classified as a VOC and is likely to be picked up in IAQ tests in homes. Benzene concentrations in the home may be the result of external air pollution coming through doors and windows and air intakes, which then becomes concentrated in fabrics and other materials. Benzene is also present in many building materials, furnishings and related products.

> Materials used in construction, remodeling and decorating are major contributors to indoor benzene concentrations. Certain furnishing materials and polymeric materials such as vinyl, PVC and rubber floorings, as well as nylon carpets and SBR-latex-backed carpets, may contain trace levels of benzene. Benzene is also present in particle board furniture, plywood, fibreglass, flooring adhesives, paints, wood paneling, caulking and paint remover. Therefore, new buildings or recently redecorated indoor environments have been associated with high concentrations of benzene from materials and furniture. Attached garages are a potential source of gasoline vapour owing to evaporation and exhaust emissions. In addition to cars, petrol, oil, paint, lacquer and hobby supplies often stored in garages can lead to increased levels of benzene indoors. Some 40–60% of benzene indoors may be attributable to the presence of an attached garage, with indoor benzene concentrations rising to 8 $\mu g/m^3$ when garages are connected to the main living environment.[43]

WHO says that 'the rate of emission of benzene from materials and furniture will decay and eventually these sources will reach a quasi-steady emission rate in new buildings within weeks or months or up to a year.'[44] However this does not explain how relatively high levels of benzene can be detected in houses some years after building work has been completed.

Toluene

Toluene will be frequently detected in IAQ tests and is reasonably anticipated to be a human carcinogen. Toluene is used in the manufacture of polyurethane foams and is widely used in furniture, mattresses and airline seats. Toluene di-isocyanate sprayed foam is widely used in homes and also steel shipping containers. The use of containers as a cheap form of housing for homeless people and refugees has to be questioned due to the cancer risks from this foam insulation.

> Toluene di-isocyanate-based rigid polyurethane foam is used in household refrigerators and for residential sheathing or commercial roofing in board or laminate form (IARC 1986). 'Pour-in-place' or 'spray-in' rigid foam is used as insulation for truck trailers, railroad freight cars, and cargo containers. Polyurethane-modified alkyds contain approximately 6% to 7% isocyanate, mostly toluene di-isocyanates, and are used as coating materials, such as floor finishes, wood finishes, and paints. Moisture-curing coatings are used as wood and concrete sealants and floor finishes.[45]

Sprayed-in polyurethane foam is very popular with many householders and businesses, as they see it as an attractive solution to insulate lofts and even walls of houses. The companies installing these foams deny there are any health risks but even so, they state that everyone should be out of the building while the foam is being installed and for at least 24 hours while the foam is curing. Emissions are much higher during the curing process.[46]

Napthalene

Napthalene will be familiar to older generations for its use in mothballs but due to its toxicity and flammability, mothballs are now made from 1,4-dichlorobenzene (which is also a suspected carcinogen). Napthalene is widely used as a plasticizer in synthetic flooring materials.[47] Napthalene has been a cause of serious emissions problems in schools and public buildings and has been restricted in Germany where flooring has had to be removed from buildings due to complaints about bad IAQ. Sometimes problems can be identified by building occupants complaining of unpleasant odours, but not all dangerous chemicals can be detected by the nose alone.

Once the problem has been detected and diagnosed there remains the issue of how to deal with it. If the chemicals are an integral part of the building construction and are trapped in the fabric of walls, floors and other parts of the building, then stripping out hazardous materials can be an expensive process. For example, a 'School in Nuremberg that had been refurbished with rubber flooring. Occupants complained of annoying smells in 2008. Indoor air quality testing revealed naphthalene concentrations from the glue used to stick down the flooring at 50 ug/m^3 whereas regulations limited naphthalene to 7–17 ug/m^3'.[48]

Napthalene has also been detected in high levels in older buildings.[49] The dangers of napthalene emissions from flooring products are well documented in Germany, USA and Canada but guidance in the UK fails to warn of indoor dangers.[50,51] There have been attempts over the years to make flooring, insulation and playground products out of recycled tyres and these have been used in the UK, despite evidence of toxic emissions, including naphthalene from such products.[52] The UK government has promoted the use of rubber crumb from recycled tyres, though mainly for external use.[53]

Tyres are significantly hazardous and not only contain benzene, naphthalene, toluene and formaldehyde but also xylene, acetaldehyde and carbon disulphide and can be responsible for very high levels of emissions. Tyre-derived products should be avoided indoors and number of studies have warned of increased danger to the population due to volatile emissions from automobile tyres.[54] An 'ecological' form of buildings, using recycled tyres, appeared some years ago called 'Earthships' and several have been built around the world including in the UK. Health risk emissions from tyres, which also include arsenic, aluminium, chromium, manganese, lead, cadmium, dioxins and furans, are strenuously denied by the proponents of Earthships and they state that emissions are restricted by plastering the tyres with earth. It is hard to understand this argument as earth is a breathable material and will not restrict the off-gassing of such a wide range of toxic emissions, but it is claimed that the plastering prevents the decomposition of tyres which could be the main cause of emissions. Scientists have also warned about using tyres for gardens as toxic chemicals will leach into water and kill entire aquatic and plant communities.[55]

Living with or excluding formaldehyde?

In Chapter 9 methods to remove, absorb or reduce formaldehyde and other carcinogenic emissions are discussed. Formaldehyde and other carcinogens discussed here are so prevalent in the environment and construction materials that it may be hard to exclude them altogether. However there are many materials that contain zero or very low levels of formaldehyde and other carcinogens and they should be selected in preference to materials with a higher risk. In Appendix A, substances that have been classified as carcinogenic or suspected of being carcinogenic are listed. It may be possible in some instances to discover from manufacturers the full chemical constituents of a product and whether it contains formaldehyde, etc. However this has become increasingly difficult. Thus while the Appendix A check list may be useful resource, it will only be so if there is a full declaration of chemical constituents.

It is worrying to find that there are so many possibly carcinogenic chemicals used in building materials and products and this should be of concern to those who want to ensure good IAQ. There will still be those who argue that there is not enough direct evidence of cancers resulting from these chemicals in buildings and this will be used as an excuse to deny there is a problem. However it should be clear from the above that it is naïve to suggest that it is possible to point to one specific product or chemical and say that will give me cancer. The problem is caused by a constant background of a cocktail of carcinogenic substances that are present in our built environment and it would be foolish to deny that this may be a contributory factor in the exponential rise in cancers in modern society. Perhaps the most convincing evidence is from people who migrate from countries where carcinogenic materials are less common and cancer levels low to developed countries where their risk of cancer is suddenly much higher.

Notes

1 Why are cancer rates increasing? Cancer Research UK http://scienceblog.cancerresearchuk.org/2015/02/04/why-are-cancer-rates-increasing/ [viewed 7.2.16]

2 How air pollution can cause cancer, Cancer Research UK www.cancerresearchuk.org/about-cancer/causes-of-cancer/air-pollution-radon-and-cancer/how-air-pollution-can-cause-cancer [viewed 7.2.16]

3 Global cancer rates could increase by 50% to 15 million by 2020, WHO www.who.int/mediacentre/news/releases/2003/pr27/en/ [viewed 7.2.16]

4 WHO (2010) Cancer prevention www.who.int/cancer/prevention/en/ [viewed 9.2.16]

5 WHO (2010) *Guidelines for Indoor Air Quality: Selected pollutants*, Geneva: WHO

6 WHO (2010) Cancer prevention, op cit.

7 Tomasetti, C. and Vogelstein, B. (2015) Variation in cancer risk among tissues can be explained by the number of stem cell divisions, *Science* www.sciencemag.org/content/347/6217/78 [viewed 7.2.16]

8 Gallagher, J. (2015) Cancer is not just 'bad luck' but down to environment, study suggests, BBC News www.bbc.co.uk/news/health-35111449 [viewed 7.2.16]

9 Stony Brook University (2015) Study reveals environment, behaviour contributes to some 80 per cent of cancers http://sb.cc.stonybrook.edu/news/general/2015-12-16-study-reveals-environment-behavior-contribute-to-some-80-percent-of-cancer.php#sthash.j1qD3mRZ.dpuf [viewed 9.2.16]

10 Wu, S., Powers, S., Zhu, W. and Hannun, Y.A. (2015) Substantial contribution of extrinsic risk factors to cancer development, *Nature*, doi:10.1038/16166

11 Israel, B. (2010) How many cancers are caused by the environment? *Scientific American* www.scientificamerican.com/article/how-many-cancers-are-caused-by-the-environment/ [viewed 10.2.16]

12 Rager, J.E., Smeester, L., Jaspers, I., Sexton, K.G. and Fry, R.C. (2011) Epigenetic changes induced by air toxics: formaldehyde exposure alters miRNA expression profiles in human lung cells, *Environmental Health Perspectives* 119 (4), 494–500. doi: 10.1289/ehp.1002614

13 Healthy Air Indoors, American Lung Association www.lung.org/associations/charters/mid-atlantic/air-quality/indoor-air-quality.html [viewed 8.2.16]

14 IARC Press Release 153, 15 June 2004 www.iarc.fr/en/media-centre/pr/2004/pr153.html [viewed 7.2.16]

15 Examples of Environmental Carcinogens, PSR www.psr.org/environment-and-health/confronting-toxics/examples-of-environmental-carcinogens.html [viewed 7.2.16]

16 www.psr.org/environment-and-health/confronting-toxics/examples-of-environmental-carcinogens.html#pbde [viewed 27.3.16]

17 Cancer and Toxic Chemicals, PSR www.psr.org/environment-and-health/confronting-toxics/cancer-and-toxic-chemicals.html [viewed 6.2.16]

18 Reuben, S.H. for Presidents Cancer Panel (2010) *Reducing Environmental Cancer Risk, What we can do now*, US Department of Health and Human Services, National Institutes of Health, National Cancer Institute

19 http://monographs.iarc.fr/ENG/Classification [viewed 3.3.16]

20 13th Report on Carcinogens, National Toxicology Programme http://ntp.niehs.nih.gov/go/roc13 [viewed 7.2.16]

21 Bizzari, S.N. (2007) *CEH Marketing Research Report: Formaldehyde*. 658.5000 A. SRI International, p. 106

22 National Toxicology Programme Board (2010) Draft Report on Carcinogens, Substance Profile for Formaldehyde (2010) Peer Review, June 21–22, National Toxicology Programme Board of Scientific Counsellors Meeting

23 Meyer, M. and Boehme, C. (1997) Formaldehyde emission from solid wood, *Forest Products Journal* 47 (5) www.ecobind.com/research/formaldehyde_emissions_from_solid_wood-mb.pdf [viewed 7.2.16]

24 Flooring Made with Composite Wood Products, Air Resources Board www.arb.ca.gov/html/fact_sheets/composite_wood_flooring_faq.pdf [viewed 10.2.16]

25 Formaldehyde Emission Standards for Composite Wood Products, EPA www2.epa.gov/formaldehyde/formaldehyde-emission-standards-composite-wood-products#proposedrule [viewed 7.2.16]

26 Formaldehyde Emissions Standards for Composite Wood Products, Federal Register www.federalregister.gov/articles/2013/06/10/2013-13258/formaldehyde-emissions-standards-for-composite-wood-products [viewed 7.2.16]

27 Formaldehyde standards for composite wood products www.gpo.gov/fdsys/pkg/BILLS-111s1660enr/pdf/BILLS-111s1660enr.pdf [viewed 7.2.16]

28 www2.dmu.dk/atmosphericenvironment/expost/database/docs/ppm_conversion.pdf [viewed 7.2.16]

29 Formaldehyde standards for composite wood products (2010) www.epa.gov/formaldehyde/formaldehyde-emission-standards-composite-wood-products [viewed 27.3.16]

30 Schwab, H. *et al.* (undated) *European Regulations for Formaldehyde*, Fraunhofer Institute for Wood Research, Wilhelm-Klauditz-Institut Braunschweig, http://owic.oregonstate.edu/sites/default/files/pubs/Schwab.pdf [viewed 23.3.16]
31 McGill, G., Oyedele, L.O. and McAllister, K. (2015) Case study investigation of indoor air quality in mechanically ventilated and naturally ventilated UK social housing, *International Journal of Sustainable Built Environment* 4, 58–77
32 Formaldehyde and Cancer risk, National Cancer Institute www.cancer.gov/about-cancer/causes-prevention/risk/substances/formaldehyde/formaldehyde-fact-sheet [viewed 7.2.16]
33 Minutes of the 401st meeting of the Northern Ireland Housing Council held on Thursday, 10 October 2013, Waterfoot Hotel, Londonderry www.nihousingcouncil.org/CMSPages/GetFile.aspx?guid=9988e9fb-6570-4318-83ec-c3fc532c57e3 [viewed 23.3.16]
34 Ross, J. (2014) *Cavity Wall Inspection Report*, South East Regional College and NIHE
35 Hoey, J.R., Turcotte, F., Couët, S. and L'Abbé, K.A. (1984) Health risks in homes insulated with urea formaldehyde foam. *Canadian Medical Association Journal*. 130 (2): 115–117
36 Urea formaldehyde foam insulation, Carson Dunlop www.carsondunlop.com/resources/articles/urea-formaldehyde-foam-insulation/ [viewed 7.2.16]
37 The Building Regulations (2010) *Approved Document D, Toxic Substances*, HM Government UK
38 Building Regulations NI (2012) *Technical Booklet B*, HM Government NI
39 Department of Communities and Local Government 2012 Consultation on Changes to the Building Regulations in England www.gov.uk/government/uploads/system/uploads/attachment_data/file/38700/2012_BR_SOR.pdf [viewed 23.3.16]
40 Health and Safety Executive www.hse.gov.uk/woodworking/faq-mdf.htm [viewed 23.3.16]
41 Vineis, P. PhD and Zeise, L. PhD (2002) Styrene-7,8-oxide and Styrene, IARC Monographs on the evaluation of carcinogenic risks to humans 82 http://monographs.iarc.fr/ENG/Publications/techrep42/TR42-9.pdf [viewed 7.2.16]
42 www.niehs.nih.gov/health/materials/styrene_508.pdf [viewed 23.3.16]
43 Harrison, R., Delgado Saborit, J.M., Dor, F. and Henderson, R. (2010) Benzene, WHO Guidelines for Indoor Air Quality: Selected Pollutants, WHO
44 Harrison *et al.* (2010) op cit.
45 National Toxicology Program, Department of Health and Human Services (2014) Report on Carcinogens, Thirteenth Edition http://ntp.niehs.nih.gov/go/roc13 [viewed 23.3.16]
46 www.icynene.com/en-us/homeowners/resources/health-safety [viewed 24.3.16]
47 Naphthalene in indoor air, Health Canada www.hc-sc.gc.ca/ewh-semt/alt_formats/pdf/pubs/air/naphthalene_fs-fi/naphthalene_fs-fi-eng.pdf [viewed 7.2.16]
48 Daumling, C., Umwelt Bundes Amt (UBA) Certech 2011. Powerpoint presentation via personal communication.
49 Lisow, W., Schmidt, M., Mertens, J., Thumulla, J., Weis, N., Köhler, M. and Pilgramm, M., Olfactometric (2015) Determination of the odour detection threshold and the identification threshold of naphthalene. Healthy Buildings Europe, Eindhoven, May
50 www.hc-sc.gc.ca/ewh-semt/pubs/air/naphthalene_fs-fi/index-eng.php [viewed 23.3.16]
51 www.gov.uk/government/uploads/system/uploads/attachment_data/file/338233/HPA_NAPHTHALENE_General_Information_v1.pdf [viewed 23.3.16]
52 California Department of Resources Recycling and Recovery (2010) *Laboratory Study Report, Tire-Derived Rubber Flooring Chemical Emissions Study*, California Department of Resources Recycling and Recovery
53 www2.wrap.org.uk/downloads/TyresRe-useRecycling.165b4623.2544.pdf [viewed 23.3.16]
54 Overton Santford, V. Manura, J.J. (1999) Volatile organic emissions from automobile tires, *Scientific Instrument Services* www.sisweb.com/referenc/applnote/app-37.htm [viewed 7.2.16]
55 www.ccathsu.com/askccat/do-earthship-tire-constructions-pose-health-risks/ [viewed 23.3.16]

5 Other hazards and radiation

Many different things affect IAQ. Hazardous pollutants come from external air pollution and from a wide range of items introduced into buildings, such as cleaning materials, perfumes, air fresheners and so on. Also a major hazard in buildings is due to the natural environment. Radon gas emanates from the ground beneath buildings and unless proper measures are taken, it can accumulate in highly dangerous concentrations causing a high risk of cancer. Some buildings can also be affected by other underground gases such as methane and emissions from old pollutants on brownfield sites where remediation has not been properly carried out. Electromagnetic fields (EMFs) can also be a serious risk to health.

While at first sight all these factors may appear to have little to do with building materials, it is important to be aware of these risks and also any connections with building materials. For instance certain forms of construction can reduce risks from EMFs and provide protection from radiation. The correct building materials and construction should be used to ensure adequate radon protection. Voids under buildings should be well ventilated to dissipate gases. However a critical issue that is often overlooked is emissions of dangerous radiation from building materials themselves. While this may not appear to be as significant as radon from the ground, radon and gamma radiation from construction materials can 'top up' radon levels in a building, pushing them over safe limits.

Gas penetration into buildings and dangers from landfill sites

In additional to radon, other kinds of gas penetration into buildings are more common than may be realised. One example is a housing development in Gorebridge, Midlothian in Scotland where high levels of CO_2 gas were discovered coming from old mine workings beneath the houses.[1] The solution was to demolish the houses! There are many areas in the UK where houses have been built over old mines. In 1988 a house in Arkwright, Derbyshire exploded as a result of methane seeping from a mine. Methane from landfill sites has also contaminated houses and there are dozens of other cases, as reported in *Inside Housing* in 2007.[2]

There are growing and disturbing reports of gas leaks into houses constructed adjacent to landfill sites. Such sites may harbour a wide range of dangerous chemicals that can find their way into houses. While this is not an issue of building materials, it can be a cause of IAQ problems, and measures to protect houses from dangerous gases, not only radon, should be an issue taken into account when renovating or designing buildings.

The most shocking example of apparent contamination took place during the 2014 UK floods in the Thames Valley in Chertsey near London when a seven year old boy, Zane Gbangbola died, and his father was left paralysed, as a result of alleged toxic chemical emissions when floodwater filled the basement of their house. Residents of neighbouring properties were evacuated due to concerns about possible toxic emissions. A campaign group claims that the death was caused by hydrogen cyanide, released due to the flooding, which was detected by the fire brigade when they attended the scene, but official bodies such as the local authority have subsequently claimed that there is no evidence of this.[3] Government statements suggested instead that the problems were caused by carbon monoxide, though it can hardly be the case that CO spread from Zane's house to the 17 other properties that were evacuated. It may be some time before the truth about this particular incident comes out as the inquest has been postponed to June 2016 and will not take place until after this book has gone to press.

> Is this the most toxic cover-up of them all? Bombshell fire service logs show they warned police of cyanide at home of boy poisoned to death in floods … So why did detectives shift blame to a faulty pump? For months after his death, police and other official agencies ruled out fears that the deadly fumes had come from a nearby landfill site.
>
> Instead, they insisted carbon monoxide from a faulty pump hired by his family had caused his death. But now The Mail on Sunday can reveal the damning evidence that proves the authorities have known for 14 months that hydrogen cyanide gas capable of killing Zane had leaked into the family's home as it was engulfed by floodwater.[4]

Paul Mobbs, writing in the *Ecologist*, has suggested that risks of contamination from landfill sites have worsened as a result of government relaxation of landfill environmental control legislation. 'Across Britain there are hundreds of sites, which have been filled with industrial or mine waste from before the 1980s. Though they may currently be stable, they may suddenly become a toxic problem as rain patterns and local hydrological systems change with the climate'.[5]

Radon – a major cause of cancer

The normal solution to avoid gas pollution in houses is to ensure that a gas-tight radon barrier is properly installed. However constructing buildings on stilts, so that air can flow underneath, is a good way to avoid radon altogether!

Guidance is available from Public Health England and other public bodies on how to deal with radon, though it focuses entirely on emissions from natural earth and bedrock.[6] Just as most buildings struggle to achieve good air tightness for thermal reasons, due to builders leaving holes or gaps in the fabric, it is possible that radon barriers are not properly installed and sealed. Service pipes can sometimes puncture the radon membrane and complicated foundation details rendering membranes ineffective. Radon sumps, to ventilate under floors, even if installed, are not always activated or are switched off by the occupants. Radon causes over 1,100 deaths per annum from lung cancer[7] but public awareness of the risks is still poor. Indoor radon is the second most important cause of lung cancer in the general population, after smoking.[8]

> About 1100 people each year die in the UK from lung cancer related to indoor radon, but current government protection policies focus mainly on the small number of homes with high radon levels and neglect the 95% of radon related deaths caused by lower levels of radon, according to a study published on bmj.com today.[9]

There is a significant failure of government to provide tests, information and guidance in a form that is understandable by the general public. Instead of carrying out a thorough survey of houses to determine radon levels, the authorities rely on educated guesswork and statistical risk analysis. This leads to a set of indicative maps that give a geographical sense of high radon risk areas but will not tell you whether your house or building may have a radon problem. You can get an estimate of the risk to your house from ukradon.org 'for a small charge' but this is not based on a test of radon levels in your property. Indicative maps are available for England, Wales, Scotland, Northern Ireland and Ireland.[10] The maps show different levels of risk but are confusing and provide a false sense of security to those where the radon maps show a low level of risk. It is possible to find high radon levels in areas of low risk.

> Although HPA [Health Protection Agency] have recently been phasing in the terms 'lower risk', 'intermediate risk' and 'higher risk' areas relating to white, yellow and brown color-coding on the maps, many people still mistakenly believe that the white areas of the map are 'not-affected' and as such are 'safe'. This is not necessarily the case, and no building can be considered to be safe from radon unless it has been tested. A white area on the map simply means that the data available does not indicate that there is a strong likelihood of radon being found there … are they really helping cut radon exposure amongst the British public or just adding to confusion?[11]

Agents, advertising property in the UK rarely, if ever, declare radon levels when they are selling houses! Thus the public is generally unaware of the risk of radon or the levels, though awareness in the USA is higher.

> Radon testing and radon mitigation are quickly becoming a common issue during real estate transactions. More and more citizens are being educated about the risks associated with radon gas. These radon-educated buyers are looking for a new home with low radon levels. Even if the buyers do not know about radon, many home inspectors offer radon testing as an option during the home inspection process and their clients choose this option often. A home with low or reduced radon levels will be more appealing to homebuyers. Homes with radon issues can be sold but homes with resolved radon issues are more sellable.[12]

Even where radon measures have been taken, post occupancy evaluation of the mitigation effects seems uncommon; it was hard to find any UK research on this though some work has been done in the USA.

> [Radon] Mitigation-system performance was adversely affected by accumulated outdoor debris blocking the outlets of subsurface pressurization pipes; fans being turned off (e.g., because of excessive noise or vibration); air-to-air heat exchanger, basement pressurization, and subsurface ventilation fans being turned off and fan speeds reduced; and crawlspace vents being closed or sealed.[13]

Radon risks from increased airtightness

There is a worry that with increasingly airtight and poorly ventilated buildings, dangerous gases may become a more serious problem. Being airtight to minimise drafts does not mean that the gas cannot get in! Various studies have shown that retrofitted houses have higher radon levels. A study of 3,300 houses in Brittany, France led to the conclusion that 'thermal retrofit could increase the indoor radon concentration significantly.'[14]

Measures to enhance energy efficiency and create buildings that are airtight raises concerns about higher levels of radon and radiation in such buildings. 'Increasing the air tightness of dwellings (without compensatory purpose-provided ventilation) increased mean indoor radon concentrations by an estimated 56.6%, from 21.2 Becquerels per cubic meter (Bq/m^3) to 33.2 Bq/m'.[15]

A study in Belgium found a radon level of 750 Bq/m^3 in a 'passive house' in a low radon risk area. This is a surprisingly high level (in the UK 200 Bq/m^3 is considered the safe limit).[16] Milner *et al.* warn of the dangers of greater airtightness in dwellings and of the risks of ventilation and mechanical ventilation and heat recovery (MVHR) systems failing, leading to a higher risk of cancer from radon, etc.

> Unless specific remediation is used, reducing the ventilation of dwellings will improve energy efficiency only at the expense of population wide adverse impact on indoor exposure to radon and risk of lung cancer. The implications of this and other consequences of changes to ventilation need to be carefully evaluated to ensure that the desirable health and environmental benefits of home energy efficiency are not compromised by avoidable negative impacts on indoor air quality … For radon at least, caution is needed to ensure that the pursuit of energy efficiency does not precipitate an unwelcome increase in disease burden in the population as a whole. It is also a reminder that all forms of mitigation action have the potential for negative as well as for positive health impacts at population level and need to be carefully planned.[17]

Radon emissions from building materials

Government guidance on radon does not refer to radon emissions from building materials and few members of the public or professionals are aware that radon is also emitted by construction materials commonly used in buildings. Steps have been taken in Europe to establish standards for measuring emissions from building materials, discussed elsewhere in this book, but this has focused on VOCs and formaldehyde. While in general terms the radiation from building materials is not high, there is evidence of significant gamma radiation from some materials in some circumstances and even where levels are low they can still top up background radiation. An EU directive now requires all member states to put plans into action so that the required radon emissions standards are achieved by February 2018.[18] The directive includes Annex (8) 'Definition and use of the activity concentration index for the gamma radiation emitted by building materials' and Annex (13) 'Indicative list of types of building materials considered with regard to their emitted gamma radiation'.

Sahoo *et al.* state that 'building materials are an important source of indoor radon, second only to soil', and have developed a methodology for measuring this. They point out that there have been many studies of radon emissions from building materials but measurements of emissions from walls in buildings have not been considered sufficiently.[19] Hoffman *et al.* are developing a VOC test chamber methodology for calculating radon emissions from construction materials.[20]

1. **Natural materials:**

   (a) Alum-shale

   (b) Building materials or additives of natural igneous origin, such as:
   - granitoides (such as granites, syenite and orthogneiss)
   - porphyries
   - tuff
   - pozzolana (pozzolanic ash)
   - lava

2. **Materials incorporating residues from industries processing naturally occurring radioactive material, such as:**
   - fly ash
   - phosphogypsum
   - phosphorus slag
   - tin slag
   - copper slag
   - red mud (residue from aluminium production)
   - residues from steel production

Figure 5.1 Indicative list of building materials with regard to gamma radiation
Source: EU Council directive 2013/59/euratom[21]

Masonry and concrete materials and radon

Materials that emit radon are largely masonry-type materials such as brick, tiles and concrete. Radon levels can vary considerably depending on the clay and rock that the products have been made from. There are also significant problems that result from the use of waste and recycled materials that are incorporated into building materials.

It would be easy to argue that it is impossible to separate radon emissions from the building fabric from naturally occurring radon in the ground or bedrock. However some interesting studies in the USA looked at radon emissions in five- and an eight-storey condominiums. Radon problems were coming from the mainly concrete construction and gypsum plasterboard in the buildings on all floors. The purpose of this study was to calculate a better ventilation system that would reduce the radon problem. Radon levels varied between 218 and 481 Bq/m^3 in *upper floors*, well above the EU/UK safe limit. Higher radon levels from the bedrock might be expected in the lower, ground floor and basement, but not in the upper floors of a multistorey building unless these were coming from the concrete and gypsum construction.[22] A study in Charlotte in the USA showed high levels of radon of three times the safe level in highrise buildings.[23]

Concrete is made from stone aggregates and these can contain uranium and radium. As radon gas decays it can produce polonium, lead and bismuth that can remain in the lungs and cause cancer. Gypsum plasterboard or drywall can also be contaminated with radioactive waste products. These materials can emit gamma radiation and radium.[24]

Exposures to gamma radiation are 40 per cent higher indoors than outdoors. They are at their lowest in countries where most people live in timber buildings and highest in countries where masonry materials are most common, thus confirming the higher risk from masonry materials. Highest emissions are from rock and stone-based materials from igneous and metamorphic rocks, such as granite. The lowest levels are from sedimentary rocks such as limestone. An exception to this is in Sweden where there are high levels of gamma radiation in housing construction from alum shale lightweight concrete (about 300,000 houses).

The highest levels of radium, thorium and potassium are found in concrete, aerated and lightweight concrete, clay (red) bricks and natural stones. The highest levels in materials made from byproducts are blast furnace slag and coal fly ash.[25]

Concrete appears to be the most significant emitter of radiation and this can vary depending on the moisture content. Radiation diminishes the dryer the concrete. Increasingly byproduct additives from fly ash and granulated blast furnace slag (GGBS) are used as cement substitute or additives in concrete. Blast furnace slag, according to Mustonen, has led to concrete with radiation emissions up to 50 per cent greater than normal concrete. Concrete products made with 15 per cent fly ash have also been shown to have slightly higher radon emanation levels.[26] Cement and concrete products that incorporate recycled waste materials are often claimed to be much more environmentally friendly than other materials but such claims do not refer to possible risks from radiation. One company that used to claim its cement product to be 'The World's most sustainable building material' was instructed by the Advertising Standards Authority of Ireland not to use such a claim in future advertising.[27]

Phosphogypsum is a waste product (from the fertilizer industry) that has been used as an asset retarder in the manufacture of Portland cement and in gypsum products. Use has been limited due to contamination with high concentrations levels of radium (200–300Bq/kg). Phosphogypsum also contains heavy metals according to the USA EPA, but despite this, it is still widely used in the manufacture of gypsum wall boards though, Kovler says this is less likely in the future due to environmental controls.[28]

Coal fly ash is a source of gamma radiation but the concrete industry argues that radiation levels in concrete mirror those in general background radiation and thus this is an issue rarely discussed when promoting the environmental benefits of reducing cement production through the use of waste substitutes. However Kovler argues, 'concrete in buildings can contribute to indoor radon levels more than any other buildings materials with the same RA content' due to the thickness and large surface of concrete in some buildings.[29]

Table 5.1 Radioactive content of building materials: radium, thorium and potassium

Material	*Becqurels per/kg (typical activity)*
Concrete ballast material	167
Concrete	174–400
Clay brick	247–670
White brick	103
Timber	2
Aerated concrete	129–430
Cement	88
Blast furnace slag	243
Byproduct gypsum	339
Natural gypsum	11
Coal fly ash	180–650
Insulation wool	39

Note: These figures are more for comparison purposes and should not be considered as absolute figures

Source: European Commission[30]

Decorative stone and granite worktops are also a source of low level radiation and have become increasingly popular in houses. Granite worktops have been found to emit around 14 Bq/kg which is relatively low, but significant enough to be of some concern.[31]

Reducing the use of concrete

Controlling emissions from construction materials is a matter for regulation, but despite EU directives, there are few regulations currently in place or enforced. Cement and concrete are some of the most widely used materials on the planet and also one of the biggest contributors to CO_2 emissions, despite recent energy efficiency improvements to cement plants. Reducing the use of concrete is a positive environmental measure for various reasons, apart from the radiation risk. The cement and concrete industry is very powerful and resists the reduced use of masonry and concrete materials even though they might reduce the contribution to radon levels in buildings. The addition of metakaolin to concrete has been shown to reduce radiation emissions by 30 per cent. Metakaolin is a highly processed, fired aluminosilicate clay-based material often used as a pozzolan with lime and in concrete.[32]

Careful consideration should be given to the use of exposed concrete and bricks on building interiors and gypsum products based on waste byproducts. Cement based on waste byproducts should also be viewed with care. Unfired earth can provide an excellent thermal mass solution alternative to cement and concrete.

Electromagnetic radiation

This is an important issue for many concerned with the health impacts of buildings on people. The German Building Biology Movement is particularly vexed with the issue of EMFs. However it is also a very large and somewhat controversial subject, which cannot be fully explored in this book, as the role of building materials is only one minor aspect. Most building materials can provide some protection from electrical fields, but not magnetic fields.

EMFs occur naturally. Light from the sun and background magnetic fields from the earth are factors. These are static fields, but buildings contain alternating currents (AC), which can be the source of another hazard. EMFs that result from ordinary household wiring should not be a risk but there has been considerable concern about the effect of microwaves from ovens and other sources.

> There are internationally accepted guidelines by the International Commission for Non-Ionising Radiation Protection ... that have been designed to limit both residential and occupational exposure to levels that are safely below those that can heat up tissue. Currently these guidelines are between 28 and 61 volts per meter (V/m) depending on frequency.[33]

External sources of EMFs are power lines and electricity substations. Often these can be located very close to houses and other buildings. Leukemia, particularly in children, is often linked to high-voltage electromagnetic radiation. 'Faulty genes makes children who live near power lines more likely to develop leukemia'.[34]

> Scientists at the Jiao Tong University School of Medicine in Shanghai studied 123 children under 15 with leukemia and found those with a faulty variant of the XRCC1 gene were 4.3 times more likely to be diagnosed with the disease if they lived within 330 feet

> of a power line or electricity transformer and the electromagnetic fields that surround such facilities.[35]

Anyone who has been near high-tension power cables, particularly during bad weather, will have experienced a strange charged atmosphere in the air and this can have negative health impacts as some consider that the highly charged air attracts air-born pollutants. Remarkably there are no restrictions in the UK on how close houses can be built to high-voltage power lines. It is not uncommon to see housing developments built right underneath power lines. However this is not allowed in the USA and in many other countries around the world where power lines have a clear green strip beneath. This can even be seen in developing countries where the residents of informal settlements avoid building underneath power lines, despite the desperate need for land. This is not a concern for housing developers in the UK, where there has been much more concern about the effect of EMFs on property values than on health.

> The presence of high-voltage overhead power lines in proximity to residential property has been the subject of periodic media attention since the mid-1980s, following the reported link between a number of adverse health effects and living in close proximity to power lines. This association remains unsubstantiated by scientific evidence and consequently no legislation exists within the UK to restrict the development of land where HVOTLs are sited. The only limitation on new development has been statutory safety clearances and as a result 'a large amount of residential development has been carried out beneath and adjacent to overhead lines.[36,37,38]

Solid construction materials can reduce the impacts of EMFs from external sources but it was difficult to find much scientific literature on this. It is well known that a Faraday cage (a metal box that shields people from electrical discharges such as lightning strikes) is one form of protection and there are examples of people incorporating metal mesh into walls to reduce the effect of EMFs. This is used in hospitals to shield exposure to MRI scanners. The use of mass concrete or metal mesh into walls may have other negative health effects however.[39,40]

For those concerned about the effect of EMFs and other forms of radiation, Building Biology Guidelines provide a great deal of useful information and set out standards and levels, some of which have been adopted by public authorities in Germany and Austria.[41]

Static electricity hazards

Some materials used in buildings such as PVC flooring can build up electrostatic charge from the air and this can have a negative effect on respiratory systems. Most people will be familiar with getting minor shocks from static build up on household and office items such as filing cabinets and even door handles. Natural materials such as wood and natural fibre carpets are electrically dissipative but many modern synthetic materials, carpets, vinyl and epoxy resin can lead to electrostatic problems.[42]

Sources of indoor pollution: cooking, heaters, furniture, cleaning materials, etc.

An important factor to bear in mind, if extensive measures have been taken to create a building with good IAQ, is that this can be undone in a matter of a few hours if the building occupants introduce a wide range of products and materials that are also sources of pollution.

In particular, cleaning fluids, biocides, aerosols, air fresheners, perfumed candles, perfumes and furniture and fittings that contain dangerous chemicals such as formaldehyde and flame retardants are all major contributors to poor IAQ. These are not explored in this book, but it is important to issue a warning. 'Why air fresheners and scented candles can wreck your health: They could cause cancerous DNA mutations and asthma: UK spends nearly £400 million a year on candles, incense and aerosols'.[43,44]

Carbon monoxide

CO is a serious hazard in buildings and is known as the 'silent killer'. Building regulations are slowly being changed to require the installation of CO detectors in houses but this only applies to new constructions. Hundreds of thousands of existing buildings do not have CO detectors. Furthermore many people do not understand the difference between smoke alarms and CO detectors. The issue achieved greater public awareness in recent times due to the tragic death of two small children, Christi and Bobby Shepherd, in 2006 in a holiday cottage in Corfu. CO can be emitted from faulty heaters and boilers, open fires, wood burning stoves, petrol appliances, camping stoves and even barbecues. Deaths have even occurred in tents from smoldering barbecues. From 1 October 2015 landlords were required to install CO detectors in rooms with solid fuel appliances in rented accommodation, though this only applies to England.[45] Different regulations apply in Scotland and Northern Ireland where CO detectors are required when new combustion appliances are installed.

Partial and uneven measures across the UK are highly unsatisfactory and much more robust and consistent policies should be applied to ensure CO protection. More information can be obtained from the Carbon Monoxide and Gas Safety Society.[46]

Notes

1 Lawrie, K., Chief Executive, Midlothian (2014) Gorebridge Gas Migration Newbyres Crescent, Gorebridge Options to Resolve Gas Migration Report Council 17.6.2014, p. 4.
2 Humphries, P. (2014) Close Encounters, *Inside Housing* www.insidehousing.co.uk/journals/insidehousing/legacydata/uploads/pdfs/IH.070216.022-024.pdf [viewed 9.2.16]
3 Truth about Zane www.truthaboutzane.com/updates [viewed 9.2.16]
4 Is this the most toxic cover-up of them all? Bombshell fire service logs show they warned police of cyanide at home of boy poisoned to death in floods … So why DID detectives shift blame to a faulty pump?, *MailOnline News*www.dailymail.co.uk/news/article-3035368/Is-toxic-cover-Bombshell-fire-service-logs-warned-police-cyanide-home-boy-poisoned-death-floods-DID-detectives-shift-blame-faulty-pump.html [viewed 9.2.16]
5 Zane: did Cameron order cover-up on landfill cyanide death of 7-year old? www.theecologist.org/News/news_analysis/2986659/zane_did_cameron_order_coverup_on_landfill_cyanide_death_of_7year_old.html [viewed 9.2.16]
6 Everything you need to know about radon, Health England www.ukradon.org/information/ [viewed 9.2.16]
7 www.ukradon.org/information/risks [viewed 9.2.16]
8 WHO (2009) *WHO Handbook on Indoor Radon: A public health perspective.* World Health Organization
9 1,000 people die a year from preventable radon leaks in home, *The Telegraph* www.telegraph.co.uk/news/health/news/4144992/1000-people-die-a-year-from-preventable-radon-leaks-in-home.html [viewed 9.2.16]
10 Miles, J.C.H. *et al.* (undated) *Indicative Atlas of Radon in England and Wales*, London: Health Protection Agency and British Geological Survey

11 Radon Maps Help or Hindrance?, Property Eco www.properteco.co.uk/radon-maps-help-or-hindrance/#sthash.FZOea0ui.dpuf [viewed 9.2.16]
12 Can I sell my home if it has high radon levels?, RadoVent www.radovent.com/Radon-Gas-Blog/bid/70381/Can-I-sell-my-home-if-it-has-high-radon-levels [viewed 9.2.16]
13 Prill, R.J., Fisk, W.J. and Turk, B.H. (1990) Evaluation of radon mitigation systems in 14 houses over a two-year period, *Journal of Air Waste Management Assoc.* 40 (5), 740–746, DOI:10.1080/10473289.1990.10466719
14 Collignan, B., Mandin, C. and Powaga, E. (2015) Impact of thermal retrofit on indoor radon exposure concentration, first results of a measurement campaign in Brittany, France, Healthy Buildings Europe Conference, Eindhoven, May
15 Milner, J., Shrubsole, C., Das, P., Ridley, I., Chalabi, Z.H., Armstrong, B. and Wilkinson, P. (2014) Home energy efficiency and radon related risk of lung cancer: modelling study, *BMJ* 348 doi: http://dx.doi.org/10.1136/bmj.f7493
16 Poffijn, A., Tonet, O., Dehandschutter, B., Roger, M. and Bouland, C. (2012) A pilot study on the air quality in passive houses with particular attention to radon http://aarst-nrpp.com/proceedings/2012/07_A_PILOT_STUDY_ON_THE_AIR_QUALITY_IN_PASSIVE_HOUSES_WITH_PARTICULAR_ATTENTION_TO_RADON.pdf [viewed 9.2.16]
17 Milner, J. *et al.* (2014) op cit.
18 The Basic Safety Standards (2013/59/EURATOM), *Official Journal of the European Union* http://eurlex.europa.eu/LexUriServ/LexUriServ.do?uri=OJ:L:2014:013:0001:0073:EN:PDF [viewed 9.2.16]
19 Sahoo, B.K., Sapra, B.K., Gaware, J.J., Kanse, S.D. and Mayya, Y.S. (2011) A model to predict radon exhalation from walls to indoor air based on the exhalation from building material samples, *Science of the Total Environment* 409, 2635–2641
20 Hofmann, M., Richter, M. and Jann, O. (2015) Determination of radon exhalation from building materials in dynamically operated test chambers by use of commercially available measuring devices, Healthy Buildings Europe Conference, Eindhoven, May
21 EU Council directive 2013/59/euratom, *Official Journal of the European Union*, 5 December 2013
22 Brodhead, B., Elevated radon levels in high-rise condominium from concrete emanation www.wpb-radon.com/pdf/Radon%20in%20high%20rise%20condominium.pdf [viewed 9.2.16]
23 Radon In Concrete, Clean Vapour LLC www.cleanvapor.com/radon-in-concrete.html [viewed 9.2.16]
24 Kovler, K. (2012) Radioactive Materials, in Pacheco-Torgal F. (ed.), *Toxicity of Building Materials*, Woodhead, Chapter 8
25 Mustonen, R. (1984) Natural radioactivity and radon exhalation from Finnish building materials, *Health Physics* 46, 1195–1203
26 Mustonen, R. (1992) *Building Materials as Sources of Indoor Exposure to Ionizing Radiation*, STUK A105, Finnish Centre for Radiation and Nuclear Safety, Helsinki
27 Advertising Standards Authority of Ireland, Case Report 21/08/2013 Ref 20593
28 Kovler (2102) op cit.
29 Kovler (2102) op cit.
30 European Commission (1999) *Radiation Protection 112, Radiological Protection Principles Concerning the Natural Radioactivity of Building Materials*, Directorate-General Environment, Nuclear Safety and Civil Protection
31 Kitto, M. *et al.* (2008) Emission of radon from decorative stone, Proceedings of the American Association of Radon Scientists and Technologists, International Symposium Las Vegas NV, AARST
32 Lau, B.M., Balendran, R.V. and Yu, K.N. (2003) Metakaolin as a radon retardant from concrete. *Radiat Prot Dosimetry* 103 (3), 273–276
33 A basic guide to EMFs, Powerwatch www.powerwatch.org.uk/science/guidetoemfs.asp [viewed 9.2.16]

34 Faulty gene makes children who live near power lines more likely to develop leukaemia (2008), MailOnline news www.dailymail.co.uk/news/article-1099077/Faulty-gene-makes-children-live-near-power-lines-likely-develop-leukaemia.html [viewed 9.2.16]
35 Chinese research links faulty gene to EMF-induced leukemia, Interference Technology www.interferencetechnology.com/chinese-research-links-faulty-gene-to-emf-induced-leukemia/ [viewed 9.2.16]
36 Sims, S. and Dent, P. (2005) High-voltage overhead power lines and property values: A residential study in the UK, *Urban Studies* 42 (4), 665–694
37 The London Hazards Centre, Fact Sheet: Electromagnetic Fields www.emfacts.com/download/gen76.pdf [viewed 9.2.16]
38 Power lines and property – UK, EMFs.info www.emfs.info/policy/property-uk/ [viewed 9.2.16]
39 Powerwatch www.powerwatch.org.uk/docs/aboutus.asp [viewed 9.2.16]
40 Maisch, D., Podd, J. and Rapley, B. (2006) *Electromagnetic Fields in the Built Environment – Design for Minimal Radiation*, Exposure BDP Environment Design Guide August 76
41 Building Biology Indoor Environment Checklist www.baubiologie.net/fileadmin/_migrated/content_uploads/VDB_Building_Biology_Indoor_Environment_Checklist__english__01.pdf [viewed 9.2.16]
42 Clements-Croome, D. (2003) *Electromagnetic Environments and Health in Building*, London: Taylor and Francis
43 Why air fresheners and scented candles can wreck your health: they could cause cancerous DNA mutations and asthma, Mail Online www.dailymail.co.uk/femail/article-3220306/Why-air-fresheners-scented-candles-wreck-health-cause-cancerous-DNA-mutations-asthma.html#ixzz3yBMdPqyx [viewed 9.2.16]
44 How is indoor air quality affected by cleaning, sanitizing and disinfecting, Fact Sheet www.epa.gov/sites/production/files/2013-08/documents/factsheet_whatisindoorairqualityandhowisitaffectedbycleaningsanitizinganddisinfecting.pdf [viewed 9.2.16]
45 www.rla.org.uk/landlord/guides/carbon-monoxide-requirements.shtml [viewed 9.2.16]
46 www.co-gassafety.co.uk [viewed 9.2.16]

6 Hazardous materials to be avoided and why

It will be clear to the readers of this book that it is not always possible to provide conclusive evidence of specific causative links between hazardous materials and health problems. Evidence is largely circumstantial and those who use, or have no fear of, hazardous chemicals, may discount much of what is included here. However, the weight of *circumstantial* evidence is significant that there are hazardous and toxic materials that create IAQ problems that can damage our health. In this chapter the intention is to draw attention to those materials that are better avoided, or at worst, used where they can have no effect on IAQ. Adopting the precautionary principle should lead to the selection of healthy alternatives that are discussed in Chapter 11. However many of the materials and products listed in this chapter are used in most building projects without a second thought, and will be found in most existing buildings. They provide the source of some of the toxic pollutants that are picked up in IAQ tests. Materials should be avoided primarily because they are carcinogens, asthmagens, endocrine disruptors, irritants or emitters of VOCs or radiation.

Treated timber

Nearly all buildings use timber somewhere. Even concrete, steel and masonry buildings use a significant amount of timber for floors, partitions, roofing and many other elements. Timber frame construction is an environmentally friendly way of building and significantly reduces the risks from radiation that are associated with masonry materials. However the industry, by and large, uses timber that has been treated with hazardous chemicals to protect against rot and insect attack. Treated timber is bad for IAQ and health. Many standard specifications and some buildings regulations require treated timber. UK building regulations require roof construction timbers to be treated against longhorn beetle in areas of Surrey and the southeast of England.

Banks and building societies frequently require a timber treatment (and damp proof) guarantee before they will lend money on a property, though a UK Building Societies Association spokesperson said they did not have any policy requiring this.[1] It seems that it is custom and practice for surveyors carrying out valuation surveys (which are not full building surveys) to recommend chemical damp proof and timber treatment, even where this may be unnecessary according to UK Damp Ltd.

> During the previous housing boom years up to 2007/8 Banks and Buildings Societies hardly worried about possible damp or timber decay problems but nowadays, even though George Osborne has manufactured a pre-election housing boom, lending criteria

> is much stricter and dampness and timber decay seem to be their number one priority when undertaking mortgage valuations or Building Surveys.[2]

Perceived dampness and timber decay problems are therefore always highlighted in a pre-purchase survey by building surveyors working on behalf of mortgage providers. The surveyor, erring on the side of caution, will generally recommend that a damp and timber report be obtained from a specialist damp proofing and timber contractor and this normally leads to the use of toxic chemical preservatives being introduced to buildings as part of a precautionary treatment against any possible infestation by woodworm, or wet and dry rot. Building surveyors are concerned to indemnify themselves against any possible negligence claims over woodworm or damp, but they are unlikely to be sued for recommending hazardous materials!

It is possible to deal with many damp, rot and insect problems without using toxic chemicals. Beetle and fungal decay are usually confined to damp timbers and can be controlled by the removal of the dampness without resorting to the use of toxic chemicals.

> Dampness in floor timbers can also be reduced by the installation of extra air bricks which will improve the flow of air under timber floors and this will purge excess moisture from the floor void, the result of this is that timbers will not be damp enough to sustain any fungal decay or insect infestation and there will be no justification for chemical woodworm or dry rot treatments.[3]

Timber treatment chemicals are mostly hazardous, though the degree of hazard is hotly debated. Chemicals used to treat timber are frequently changed and full details are rarely disclosed so it can be difficult to identify the degree of hazard. However, emissions from timber treatment will undoubtedly contribute to poor IAQ.

Some timber is enclosed within impermeable building structures, isolated from indoor air (in theory) so that the chemicals used are unlikely to affect IAQ, but a lot of treated timber is quite close to occupants and exposed within rooms, roof spaces or dry lining, covered only by gypsum plasterboard and other finishing materials. It is normal practice to use treated timber in roofs as these are seen as at greater risk of damp.

Extreme examples of danger are where timber has been treated with banned chemicals such as creosote and copper chrome arsenic (CCA). Some building projects may use recycled timber that could have been treated in the past with hazardous substances. The chemicals can take many years to dissipate. Much of the literature on hazardous timber treatment stems from the 1980s and 1990s and it has become less of a hot topic in recent years as the industry had to remove many toxic materials from the market, though this does not mean that the hazards are no longer there.

The London Hazards Centre published a 200-page report in 1989 that challenged the wood preserving industry's claim that spraying pesticides in homes was safe. The London Hazards Centre also documented the deaths of UK employees of wood preservative companies of leukaemia. In Germany, the Association of Victims of Timber Treatment had 5,500 members.

> Chemicals, which are not allowed on our fields (such as lindane), may be sprayed freely in our homes and places of work. Three million houses have been treated already. Next year at least half a million people including children, babies and the yet unborn, will receive what for many will be the biggest pesticide dose of their lives.[4]

While attempts have been made to remove these highly dangerous toxic pesticide treatments from homes, it is almost impossible to do so, and such chemicals linger in our indoor environment as a result of timber treatments over several decades. Pesticides and timber treatment chemicals can be absorbed by our bodies and stay there for a long time. A report commissioned by the World Wide Fund for Nature (WWF) in 2003 found high levels of banned pesticides in the blood of young people (WWF employees) and high levels of flame retardant chemicals (PBDEs) way above generally accepted safe levels. The highest levels of pesticides found in the blood of volunteers was in Belfast, though the individual with the highest lindane contamination was in Exeter.[5] Lindane was one of the highly toxic chemicals identified by Rachel Carson in her hugely important book *Silent Spring*. Despite Carson warning us of the dangers of pesticides in 1962, the pesticides industry has grown substantially since then and remains a massive problem for health as reported in *Silent Spring Revisited*.

> This report exposes the (American) EPA's failure to protect people and wildlife from exposure to harmful pesticides and highlights the agency's on-going refusal to reform pesticide use in accordance with scientific findings. It reveals what the EPA ignores: Pesticides pollute significant areas of our air and water, threaten endangered species, and continuously expose farm workers, women of reproductive age and children to harmful levels of chemicals.[6]

Chemicals used to treat timber

Timber, used in buildings, is commonly pressure treated with hazardous chemicals. This is an issue for new building construction as well as existing buildings. Timber treatment is often referred to as 'tanalised' from the term 'tanalin' and tanalising has become the commonplace name for any sort of timber treatment, though Tanalised is a trademarked name and no inference or criticism of this particular product should be drawn from the following. There are numerous other trade names for different products from different companies.

Creosote

Creosote has been banned from sale to domestic customers for DIY (do-it-yourself) use since 2003 and its use has to be authorised since 2013, however it is still widely used.[7] It has been substituted by chemicals claimed to be less harmful, mainly for external use, but some creosote is still permitted in certain industrial applications, such as telegraph poles. The replacements have a brown stain in an effort to make them look like the original stuff!

Arsenic

Arsenic has been used in pesticides and timber treatment. It is listed as a carcinogen. CCA began to be used for timber treatment in the 1930s but has subsequently been banned or at least severely restricted in many countries from about 2003. Some industrial companies still claim that CCA is not dangerous but it has largely been replaced by substitutes that are claimed to be less hazardous. In addition to the use of arsenic in CCA, chromium is also hazardous and has been listed as a human carcinogen.

Copper azole

There are a number of so-called new generation timber treatments following the banning of CCA.[8] Copper azole is claimed to be less dangerous though it is still classified on data sheets as hazardous. Tanalith E data sheet lists the following chemical constituents: copper carbonate, 2-aminoethanol, boric acid, tebuconazole, propiconazole, polyethyleneamine, organic acid and surfactant. 'Bayer Chemicals, which makes the azole component of the new treatment, commissioned the health risk assessment, which was conducted by Gradient Corp. That study concluded that "no adverse health effects are expected"'.[9]

Pentachlorophenol

Pentachlorophenol (PCP) is a chlorinated hydrocarbon insecticide and fungicide used to protect timber from fungal rot and wood-boring insects. PCP is also used as a herbicide.[10] PCPs are banned in many countries, apart from the USA, following the Stockholm Convention on Persistent Organic Pollutants in 2001, defined as 'chemical substances that persist in the environment, bio-accumulate through the food web, and pose a risk of causing adverse effects to human health and the environment'.[11]

Naphtha (petroleum)

Low-boiling point hydrogen treated naphtha is also used in timber treatment as a fungicide and insecticide. It is considered hazardous by the Occupational Safety and Health Administration (OSHA) and listed in health and safety data sheets as extremely flammable, irritating to eyes and the respiratory system. It affects the central nervous system and is harmful or fatal if swallowed.[12]

Permethrin

Permethrin is commonly used as a 'safer' alternative to CCA to deal with woodworm but is also possibly toxic. Several product data sheets for permethrin describe it as including propynyl-butyl carbamate with propiconazole. This is a water-based preservative and biocide used in wood treatment and paints as well as cosmetics! However, it has been described as a contact allergen, which might cause skin problems.[13]

A 'woodworm expert', on a web advice site, also says they are safe whilst warning you to keep your fish tank covered whilst work is carried out!

> Q. How safe are modern woodworm treatments?
>
> A. The simple answer is – very safe. The UK HSE licensed insecticides have all been tested over many years and have a good safety record for humans in general. However, there are areas of concern, so before using ANY CHEMICAL in your house you need to consider a few issues … some chemicals in common use are highly toxic to fish and birds (Pyrethroids – Permethrin, Cypermethrin, Deltamethrin). Remember that the insecticide chemical may be 'carried' in other chemicals – emulsions or solvents for example – these can leave an unpleasant odour. Thirdly, does the proposed chemical give off any vapour – so will it leave the wood over time – could it cause problems for sensitive people in your house, like babies or asthmatics?[14]

Does timber need to be treated?

The timber treatment industry is very cautious about discussing its use of dangerous chemicals. You will struggle to find any reference to chemicals on the industry's websites where they simply state that all chemicals used are safe.[15]

It would have been impractical to list in this book all the timber treatments and preservatives, as there are many. They are increasingly water, rather than solvent, based and it is increasingly difficult to discover their chemical composition, as this is often not listed on the container or the technical data sheet. Since the EU REACH directive, companies are able to avoid providing a full chemical and risk analysis on their data sheets by simply declaring 'All constituents of this product are listed in EINECS (European Inventory of Existing Commercial Chemical Substances) or ELNCS (European List of New Chemical Substances) or are exempt' or 'Not regarded as a health or environmental hazard under current legislation'.[16]

It is common practice for architects to specify treated softwood timber and as much of the carcassing timber used in buildings comes from poor quality sapwood there is always a small element of risk of rot or insect infestation. Indeed rot can occur in treated timber if it is constantly exposed to dampness. There is a growing view that not all timber needs to be treated. If exposed to the air it can be protected with environmentally friendly stains and paints and if inside a building, care can be taken to avoid the need for timber treatment.

Joiners and carpenters are at a particularly high risk from dust in workshops when machining timber if it has been treated as the toxic chemicals are released in higher doses.[17,18,19,20] There are also reported cases of people becoming ill from toxic chemicals released from scrap wood burnt in open fires and on barbecues.

Flame retardants

Chemicals are also used with timber as flame retardants and these are discussed in more detail below. In addition to the hazardous chemicals used, both flame retardants and pesticides have for many years been delivered through solvent chemical carriers, though, as the solvents are dangerous to health, increasingly solvents are being replaced by water.

Borax

Sodium borate or borax, borates or boric acid have been used in timber preservatives and flame retardants. Borax is commonly used in cellulose and some sheep wool insulation. The manufacturers claim that it has a very low toxicity but this is a controversial claim.

Borax has been popular with many people looking for an alternative to more toxic chemical treatments and is widely used in a range of timber and natural insulation products. However, borax is water soluble and will not be effective for outdoor use. Boron, borax, borates, etc. are available in a wide range of formulations such as sodium tetraborate decahydrate. There has been considerable debate about the toxicity of borax and in 2010, it was added to the REACH Directive candidate list as potentially harmful to human reproduction. However many argue that as borax is widely present in nature it should not be seen as harmful in products.

> We've been using boric acid and borate compounds in products for a long time – in cosmetics, cleaning products (the laundry staple, Borax), as a 'non-toxic' pesticide and as an ingredient in building products. You find it in cellulose insulation, bamboo

> treatments, mattresses, termite protection systems, treated lumber, paints and coatings, and elsewhere. Furthermore, boric acid is a vital component to cellular structure in plants and a limiting micronutrient in agriculture. That means we spread it around on fields and elsewhere. We're eating it all the time and we have been for years. Our resulting policy for GreenSpec is as follows: Don't worry about low concentrations of borate compounds in products. It's simply not a problem.[21]

However, GreenSpec warn that it is important to scrutinise products such as insulation. For instance, borates can be 20 per cent of the weight of cellulose insulation and it is also widely used in some sheep wool insulations where borax dust can be a problem. Installers of some sheep wool products that are heavily dosed with borax complain of eye infections and asthmatic reactions when installing these products. Not all sheep wool insulations use these compounds however.

Literature from an American cellulose insulation product states up to 10 per cent boric acid, up to 11 per cent ammonium sulphate and up to 2 per cent zinc sulphate in its ingredients but not less than 85 per cent newsprint. Some cellulose insulation products made in Europe use magnesium sulphate (Epsom salts) up to 15 per cent and 5–6 per cent boric acid. Other European cellulose insulation products use 9 per cent aluminium hydroxide with boric acid contents going as high as 25 per cent, but normally it is around 6 per cent. However there is a Natureplus certified cellulose insulation product that claims to be borax and ammonia free.[22]

Rio Tinto minerals is the main producer of borax. 'Rio Tinto Minerals supplies nearly half the global demand for refined borates, key ingredients in fiberglass, glass, ceramics, fertilizers, wood preservatives and hundreds of other products'.[23] However, Rio Tinto does not have a spotless record in terms of environmental compliance.

> The British mining company that provided the gold, silver and bronze for London 2012 Olympic medals is being sued over claims of 'illegal air pollution' at the open pit mine that produced the Olympic metals. Doctors, environmentalists and concerned citizens are suing the London-based mining company Rio Tinto in a case that starts on Tuesday. They allege the £60bn company violated clean air laws at the US mine that produced 99% of the metals used in the 4,700 Olympic and Paralympic medals.
>
> The lawsuit filed at Utah district court against Rio subsidiary Kennecott claims the Bingham Canyon mine, near Salt Lake City, has breached air pollution laws for five years causing effects doctors called 'similar to smoking 20 cigarettes a day'[24]

The Eden Project in Cornwall makes extravagant claims as to the environmental qualities of its Education Centre building which has a copper roof. The copper for the roof comes from the same Kennecott mine in Utah where there has been a history of environmental prosecutions.

> The copper for the roof was donated by Rio Tinto and sourced from Kennecott Utah Copper Company's Bingham Canyon mine, near Salt Lake City, Utah, USA – regarded as being among the most responsibly managed copper mines in the world. One of the features of operations at Bingham Canyon has been a progressive clean-up and regeneration. In recent years, Kennecott has spent around US$350 million on the clean-up of historic smelter operations. Since 1994, it has reclaimed more than 16,800 acres of land for wildlife and has planted more than 120,000 trees over the same period.[25]

PBDE flame retardants

PBDEs have been widely used as flame retardants in a wide range of building, domestic and electronic office products and polyurethane foams. PBDE chemicals were found in the blood of many of the WWF volunteers referred to in the 2003 Contamination Report.[26] In theory the use of PBDEs is meant to have been discontinued and other chemicals have been introduced, some of them phosphorous based.

Brominated flame retardants (HBCD – hexabromocyclododecane) have been added to polystyrene insulations. 27,000 tonnes of HBCD were manufactured in 2012. HCBD is an endocrine disruptor and bio-accumulates in animal fat, becoming more and more concentrated as it moves up the food chain. Concentrations are highest in seals, dolphins and whales and it is ingested in high quantities by Inuit people. Chlorinated flame retardant TCPP (tris 1-chloro-2 phosphate) is used in polyurethane and polyisocyanurate insulation and is a potential carcinogen.[27]

The flame retardants that are bromine or phosphorous based represent some of the most serious and persistent pollutants found both in buildings and the wider environment as they have a tendency to accumulate both in materials and our bodies.[28]

> According to Dr. Heather Stapleton, the use of one flame-retardant mixture, known as PentaBDE (PBDE), was phased out in 2004, due to concerns about its tendency to concentrate in human tissues and lead to potential human health effects. Other chemicals are currently used to meet flammability requirements, but little information is available on how people are exposed to these new flame retardants or their potential health effects.[29]

Various studies have shown that alternatives such as flame-retardant mixtures based on triphenyl phosphate (TPP) came into use as a replacement for the phased-out PBDE. These showed that perinatal exposure to TPP is associated with endocrine-disrupting effects.[30] TPP is described in another study as 'ubiquitous in house dust' and it may be 'obesogenic', leading to obesity.[31] In an effort to reduce the use of toxic flame retardants, new chemical products have been introduced and widely used in products and buildings before the possible health impacts have been understood. Halogenated phenols like PCP, 2, 4, 6-tribromophenol, triclosan and tetrabromobisphenol A are commercially used as pesticides, wood preservatives, antimicrobials and flame retardants. These chemicals can persist in the indoor environment and can accumulate in human blood and tissues.[32]

Flame retardants in timber

As referred to above, chemicals are widely used to treat timber to protect against insect attack and rot, but there is also an irrational fear of timber as a fire hazard. It is possible to specify untreated timber if the building occupant wishes to avoid risks of chemical emissions. Building control officers often insist on the use of fire retardant paints on internal timber even though the requirement in the regulations is only that a building will remain structurally stable for a 'reasonable period'. Regulations to limit surface spread of flame may often be used against timber but there are many issues with how combustibility is interpreted. There are many other materials used on surfaces in buildings that are far more combustible than timber and it is questionable how effective flame retardants actually are in dealing with surface spread of flame.

Timber in relatively big sections is safe in fires as wood chars on the outside. In the event of a fire there are also other materials present inside buildings that burn much more quickly, though once a fire becomes established, smaller sections of timber will of course also burn.

The requirement for fire retardant chemicals in a wide range of products has been questioned due to their toxicity and also as they are relatively ineffective. Strong arguments have been made for their removal from building codes in the USA.[33] 'Flame retardants whose primary use is in building insulation are found at increasing levels in household dust, human body fluids and in the environment. These substances have been associated with neurological and developmental toxicity, endocrine disruption, and potential carcinogenicity'.[34]

> All foam plastic building insulation in the United States contains flame-retardant chemicals that are known to be persistent and harmful to health or lack adequate toxicity information. The use of these insulation materials (polystyrene, polyurethane, and polyisocyanurate) is increasing as buildings become better insulated and more energy efficient. People are exposed to the flame retardants used in such insulation … These exposures represent a significant occupational and environmental hazard. Studies suggest that the use of flame retardants in foam plastic building insulation is not needed to ensure the fire safety of buildings, and building codes in Sweden and Norway have been updated to allow the safe use of insulation materials without flame retardants. This would greatly reduce the use of flame-retardant chemicals and their potential for human health and environmental harms.[35]

> As evidence linking the use of halogenated flame retardants to health risks continues to mount, there is increasing pressure on government and industry to take action. About a dozen U.S. states have enacted laws that bar certain uses of various flame retardants. Among these regulations are those that bar the use of two or more polybrominated diphenyl ethers. Flame retardants are still found in products from which they have been barred, probably due to poor oversight of supply chains (PBDEs), particularly in children's products. New York recently passed a law limiting use of the flame retardant known as Tris, while the European Union limits the use of certain halogenated flame retardants in electronics.[36]

Adhesives and sealants

Composite timber products are widely used in buildings and the risks from adhesives containing formaldehyde and other chemicals are referred to in Chapter 4. Adhesives and sealants are used in much greater quantities in buildings today than was the case a few decades ago. Buildings are glued together and airtightness boards and membranes are sealed up with synthetic adhesive tape. Little thought is given to the health risks of the many materials that are used both for building workers and subsequent emissions into buildings from these materials. In Germany, the industry has recognised that there are problems and guidelines have been developed.

> In Germany the Responsible Care and Sustainable Development Guidelines of the VCI (Verband der Chemischen Industrie) have governed the development and manufacture of adhesives. This specifically means that health protection and environmental compatibility considerations are taken into account when developing and manufacturing new products. This has consequences for the composition of the adhesives and sealants, product design, recommendations for the application of the adhesive or sealant, and the purpose of use and for the recycling of the product after it has been used.[37]

A particular concern in terms of sealants is polychlorinated biphenyl (PCB) contamination. PCBs are classified as carcinogenic and have caused serious pollution concerns as they leach out into the wider environment. Some research has been done, not so much on indoor contamination but on how PCBs are found in residential areas.[38] PCBs were commonly used in polysulphide 'caulking' materials and the American EPA identified this as a health risk in school buildings in 2015.[39] The EPA identified PCBs in caulking, paints, coatings and other materials. Caulking is widely used to seal gaps around doors and windows and in wet areas such as bathrooms and toilets. PCBs were used in these sealants up to the 1970s when they were banned but they have been found in worrying levels in buildings up to today. Sealants are also used externally to seal gaps around panel materials and gradually broke down due to weathering, leaching contamination into the surrounding soil.

Modern sealants and adhesives are made from a wide range of materials, many of which may be hazardous, including acrylics, neoprene, polysulfides, polyurethanes, silicone and vinyls and epoxys, etc. Some caulks and sealants are marketed as healthy and good for IAQ but this should not be taken at face value.

It is not uncommon for people applying sealants to lick their fingers when trying to achieve a smooth finish and transfer possible contamination directly to their mouths. Many sealants and adhesives are applied with the use of protective gloves and masks.

> Silicone caulk can adversely affect your body if you swallow it or inhale it. Caulk also has toxic effects if you absorb it through your skin, according to the warning labels on the tubes of this product. This means that if you use your hands to spread this silicone caulking material, the toxicity may absorb through the surface of your skin and into your body. If this occurs, physical reactions include irritation to skin, eyes and respiratory system. It can also cause general nervous system depression.[40]

It is interesting to find, when researching the manufacture of silicone and polymer sealants and adhesives, that the same chemical company names appear that are responsible for other hazardous materials referred to in this book. Much of the health and safety advice available online is generated by the manufacturers. Adhesives and sealants may have a negative effect on IAQ due to emissions during initial curing, though insufficient information on this is currently available. If alternatives are available that avoid the use of synthetic products such as mechanical fixing and natural materials then these should be preferred. Simply assuming that products that are currently in use are safe and acceptable, when so little information is available on hazards, would be misguided.

Asbestos, fibreglass and other related materials

Asbestos

It would be highly unlikely for asbestos products to be specified for work in any building today as it is illegal in Europe, and yet asbestos is still mined and produced in some parts of the world such as Russia, China, Brazil and Kazakhstan and surprisingly Canada, where 100,000 tonnes were produced in 2010. Russia produces a million tonnes of asbestos per annum and appears to be the main consumer of asbestos products.[41,42]

Asbestos is still present in many existing buildings and efforts to remove it are not moving fast enough, according to some reports. Ironically, a leading cancer hospital, the Christie in

Manchester, has been involved in a dispute with workers as they claim to have been exposed to asbestos that had been known about for over ten years.[43]

> Thirteen people a day in the UK die from exposure to asbestos – more than double the number that die on the roads. In the USA, asbestos will be responsible for around 10,000 deaths this year, meaning it kills close to as many people as gun crime or skin cancer.[44]

An all-party UK parliamentary group report on asbestos published in October 2015 stated that at least 5,000 people die annually from exposure to asbestos.[45] There is an assumption that asbestos is mainly a problem that affects workers involved in removing it, but deaths from mesothelioma affect people right across the population. Official statistics may not record all of these deaths, as the diagnosis may not always be apparent. The fact that deaths related to asbestos exposure are *increasing* is a disturbing fact. A report by the European Trade Union Institute estimates over 47,000 people per annum in the EU are dying of asbestos-related conditions.[46]

A schoolteacher, Elizabeth Bell, in Lincolnshire, UK, who worked in schools in the county from 1968 to 1995, died in 2015 from the asbestos-related cancer mesothelioma, as recorded by the coroner. The cause was claimed to be from asbestos in display boards. Reports claim that 75 per cent of UK schools still contain asbestos and that funding cuts to the Health and Safety Executive mean that risks to school employees are greater.[47]

Given the well-documented dangers of asbestos it is not proposed to devote much more space to it in this book, but it is important to be aware of the various measures that can be taken. In some cases it may be sensible to clean and seal existing asbestos cement roofing materials for instance. Removal and demolition can be more dangerous as fibres may be released to the atmosphere. Sealing in asbestos may be the answer in some circumstances, though it was difficult to find any independent literature recording the success of this.[48] However asbestos remains a serious problem in buildings and should be checked in every renovation project. As deaths from asbestos are rising rather than falling, this will be a serious problems for many years to come.

Mineral and glass fibres

Mineral and glass fibre insulation products are preferred by many architects and their clients to foam insulation products, as it is assumed that they are safer in fires and are more environmentally friendly, as some are claimed to incorporate recycled materials. While certain mineral fibre products are fire resistant, claims about recycled content can be exaggerated. The UK Advertising Standards Authority in 2011 said that the Mineral Wool Insulation Manufacturers Association (MIMA) did not provide any solid evidence of recycling. MIMA are part of EURIMA.[49]

There are serious concerns about adhesives, resins and additives used in mineral fibres, plus concerns about fibres released into the air that can have a negative effect on IAQ. Mineral fibres are products spun from either glass or melted stone. Insulating products are often referred to as 'glass wool' or simply mineral fibre by the manufacturers in an effort to make them sound more attractive, as many people are unwilling to handle mineral fibres as they believe they make their skin itchy. Changing the name does not remove these concerns! Scientific research has led to concern that glass wool fibres, if inhaled, can lead to cancer.

> Glass wool fibres (inhalable) are reasonably anticipated to be human carcinogens based on sufficient evidence of carcinogenicity from studies in experimental animals of inhalable glass wool fibers as a class (defined below) and evidence from studies of fiber properties which indicates that only certain fibers within this class specifically, fibers that are biopersistent in the lung or tracheobronchial region are reasonably anticipated to be human carcinogens.
>
> Because there is considerable variation in the physicochemical and biophysical properties of individual glass wool fibers, carcinogenic potential must be assessed on a case-by-case basis in experimental animals, through either long-term carcinogenicity assays measuring the persistence of fibers in the lung.[50]

There has been considerable controversy about whether glass fibres are carcinogenic, but for a time, in the USA, they were classified as suspected carcinogens and warning labels were carried on packaging.

> Fibreglass is now causing serious health concerns among US officials and health researchers. In a series of papers published from 1969 to 1977, Dr Mearl F. Stanton of the National Cancer Institute found that glass fibres less then 3 microns in diameter and greater than 20 microns in length are 'potent carcinogens' in rats; and, he said in 1974, 'it is unlikely that different mechanisms are operative in man'. The International Agency for Research on Cancer (IARC) of the World Health Organization listed fibreglass as a 'probable [human] carcinogen' in 1987. In 1990, the members of the US National Toxicology Program (NTP) – representatives of 10 federal health agencies – concluded unanimously that fibreglass 'may reasonably be anticipated to be a carcinogen' in humans.[51]

> Chronic Obstructive Pulmonary Disorder (COPD), colon cancer, ulcers and lymphomas may all be caused by attenuated fiberglass. According to the American Lung Association, fiberglass insulation packages display cancer warning labels.'These labels,' says the Lung Association, 'are required by the U.S. Occupational Safety and Health Administration (OSHA) based on determinations made by the International Agency for Research on Cancer and the National Toxicology Program'.[52]

However, by 2011, after a massive campaign by the manufacturers, public health bodies such as the National Toxicology Programme (NTP) in the USA were forced to withdraw the carcinogenic classification. The fibreglass industry has been very active in challenging much of the scientific work on health risks. The length and size of fibres is one area of dispute as the fibres are not as microscopic as asbestos and it is claimed by the industry that they can be expelled by the human body. However, health concerns are not just limited to cancer. Many building workers refuse to handle glass fibre insulation due to concerns about itchiness on the skin and skin rashes. 'Aside from the cancer risks, exposure to fiberglass is known to cause: Irritation to eyes, nose and throat, rash and itchiness, stomach irritation if fibres are swallowed, worsening of asthma and bronchitis'.[53]

In addition to concerns about fibres, the use of formaldehyde (as discussed in Chapter 4) has been an issue for mineral fibre insulation. Phenol formaldehyde resins and catalyst coatings are added to non-woven fibre matt, which is then compressed and heated in an oven.[54,55]

Fibreglass dust can be generated by this process, which remains throughout its life and thus houses can be contaminated with this dust, and the dust may be contaminated with formaldehyde from some products. Many fiberglass and some stone wool fibre products were made for years using a formaldehyde-based glue and these remain in thousands of houses in the UK and elsewhere.[56]

Due to concerns about formaldehyde, manufacturers began to develop various new binders, using compounds from urea, melamine, dicyandmide and methylol. Natural oils such as linseed, soy bean, fish or lard oil have also been tried. One leading fiberglass manufacturer began to promote an alternative 'brown' fiberglass product as free of formaldehyde, even though this did not apply to some of its other products.

As a result of concerns about formaldehyde some, if not all, UK and US manufacturers have changed the constituents of their fiberglass, claiming that there is either no or low formaldehyde content. 'As of October 2015, every fibreglass insulation company in the United States and Canada has phased out the use of formaldehyde-based binders in lightweight residential products'.[57] For instance one company says that phenol formaldehyde has been replaced by a 'biolatexpolymer' binder that is claimed to be 100 per cent natural and free of VOCs in some products. It states that the colour of insulations comes from the binder that is used, such as yellow or greeny grey but their new phenol/formaldehyde free product is brown.[58]

It is difficult in the UK to find out whether products are phenol formaldehyde and amine free, as manufacturers do not disclose the constituents of their products or state VOC emissions. However rival companies have fallen out with each other over the formaldehyde-free claims.

> As you may have heard, a fibreglass insulation producer is now manufacturing and selling a product that they promote as being 'formaldehyde-free.' Many customers have asked us about formaldehyde in fibreglass insulation and what we think about our competitor's product ... As the leading producer of fibreglass insulation, we want to address that confusion and possible concern, and set the record straight regarding formaldehyde and fibreglass insulation.[59]

Owens Corning claim that all fibreglass products contain 'trace elements' of formaldehyde but that this is not dangerous and quote their own trade association to back this up. 'Consistent with the Environmental Protection Agency and the U.S. Consumer Product Safety Commission, we do not consider the trace amounts of formaldehyde found in fibreglass insulation to be a concern to human health or the environment.'[60]

A leading environmental testing organisation, Envirotest Ltd., points out that studies about the health hazards of fibreglass come to widely varying conclusions. It encourages readers of studies to examine carefully the methodology used and who has financed the studies![61]

Why are foam insulations dangerous?

There is a universal assumption that all insulation is a good thing, as it will make buildings and homes warmer and healthier. In my book *Low Impact Building*, I explained how for many architects and specifiers, 'insulation is insulation is insulation', without any distinction as to type or performance.[62] Much literature on improving the thermal efficiency of buildings simply refers to thicknesses of insulation without specifying what kind of insulation. In practice different kinds of insulation perform very differently – some perform more effectively than others, some are airtight, some are vapour permeable, some have thermal mass and others do not. Recent studies about the so-called performance gap explain that many insulation materials rarely achieve the thermal resistance claimed by manufacturers and often how well insulations are installed can make a huge difference to their effectiveness.[63]

What is rarely discussed is that the bulk of foam insulation materials are synthesised using a wide range of chemicals or natural materials combined with chemicals. Most of these chemicals are hazardous during manufacture and for the environment, when in buildings, when they catch fire and when disposed off at the end of life. Despite the assumption that insulating buildings is good for the planet by saving energy, insulation materials should also be regarded as some of the most environmentally damaging products.

Foam insulation materials contain hazardous chemicals that can emit VOCs and carcinogens. The hazardous chemicals are in a cocktail of petrochemicals, oils, binders, adhesives, solvents, blowing agents, flame retardants, and stabilisers. Not all of these are included in every product, but most are hazardous either individually or when combined.

A certain amount of complacency about hazards and toxicity of insulation exists in the industry as it is assumed that most insulation is sealed into cavity walls, under floors or in roofs where it is unlikely to have much, if any, effect on IAQ. Slabs or rolls of insulation appear as an inert and innocent standard item on building sites. Manufacturers of insulation products will deny that there are any health risks from their products and that any hazardous off-gassing happens very early in the life of the product, quickly dissipates and is no longer a health risk.

Occupational risks in manufacture

There is no doubt that serious risks exist for those workers involved in manufacturing synthetic insulation materials. Having visited several insulation factories, I have seen a wide range of health and safety practices. In a closed cell polyurethane foam insulation factory, the serious chemical interactions took place in a sealed chamber, where presumably workers only had access for cleaning and maintenance. Few odours were obvious except in the storage warehouse where tens of thousands of sheets of insulation were stacked. In another factory, manufacturing expanded polystyrene, the smell of the blowing agent pentane was so overwhelming that it made me nauseous.

Pentane and isopentane (2-methylbutane) have been widely used as a blowing agent in the manufacture of polystyrene insulation but this is being replaced by CO_2 in some products. Pentane commonly appears as a VOC in IAQ reports. Pentane is a VOC that was introduced as a blowing agent to replace banned hydrofluorocarbons (HFC). It has a strong characteristic sweet smell and evaporates easily. It is highly flammable and can be explosive when mixed with air. It is claimed to be 'safe' when emissions are at low levels. 'Inhalation of air containing high levels of pentane can cause irritation of the respiratory tract, drowsiness, headache, dizziness, a burning sensation in the chest, unconsciousness and in extreme cases coma and death'.[64] 'High-efficiency carbon dioxide blowing agents for foaming can reduce or eliminate the need for flammable blowing agents, such as pentane and butane and other ozone-depleting compounds'.[65]

Some expanded polystyrene manufacturers use a chemical blowing agent that is claimed to be more effective than CO_2. Known as trans-1-chloro-3, 3, 3-trifluoropropene, it is a liquid halogenated olefin. The manufacturer has set an occupational exposure limit of 800 ppm for this gas.[66]

It takes persistence and considerable detective skills to discover what chemicals are used in foamed insulation, base materials, foaming agents and fire retardants. Even less clear is the evidence of health problems associated with these chemicals. However any of the hazardous chemicals in synthetic insulation manufacture are linked with a wide range of occupational illnesses including cancer, asthma and respiratory conditions. These hazards can be serious for installers of insulation materials when they are handling them on a daily basis. It is not uncommon to see workers installing insulation into lofts without wearing any protective gloves or facemasks.

Given the lack of disclosure and concerns about the wide spread of chemicals involved in synthetic insulations it would be best to apply the precautionary principle and avoid their use if there is any risk of indoor air contamination from such materials.

It is important to assess the risks to occupants of buildings and their degree of exposure to VOCs and fibres released by insulation materials. Often the only separation between the insulation and the interior is a thin sheet of plasterboard in a ceiling or wall. Increasingly insulation is being installed on the inside of rooms. There are many composite wall products where foam insulation is bonded to the plasterboard and other materials using adhesives that may also be hazardous.

Fire risks of foam insulation

Hazardous insulation placed on the outside of walls or in sealed cavities should not damage the indoor environment, however, even on the outside of buildings, insulation can be a serious health risk in fires. There have been a number of serious fires in buildings that have been over-clad with synthetic insulation, which when it burns, releases highly dangerous gases such as hydrogen cyanide and CO, which can kill within seconds.

It was suggested at an inquest that this may have contributed to the deaths of six people in Lakanal House in southeast London in July 2009, as the block had been retrofitted with new insulated cladding panels and the fire resistant properties of these were questioned at the inquest.[67,68] Serious fires in several multistorey hotel and apartment buildings in Dubai, recently, have been attributed to synthetic insulation in the external cladding.[69]

Rather surprisingly, the most serious toxic emissions in fires are from the flame-retardant chemicals than are used to try to prevent fires. Flame retardants, as discussed above, are among the most serious sources of hazardous chemicals that damage IAQ. Not only are flame retardants dangerous in fires but they also represent one of the most significant sources of toxic chemical emissions in buildings.

It is difficult to find the regulations that specify or control the use of flame retardants in insulation. UK buildings regulations lay down a range of standards for non-combustibility and presumably insulation materials are assumed to comply with these. However most synthetic insulation materials (apart from mineral fibre and natural insulations) are far from non-combustible. Regulations exist for furniture, mattresses and children's nightwear and there have been significant changes to these in recent years as the toxicity of flame retardants was realised, but much remains to be done to clarify risks in building insulation materials.

Polyurethane spray foam

Sprayed polyurethane insulation has become popular in the UK and Ireland, particularly for roofs and cavities. Extravagant claims are made for its insulation properties. Some brands have incorporated a small amount of soya additives, leading to claims that it is an ecological product, but the foam is still largely based on urethane. In the USA there have been many

Polyurethane rigid foam (PUR)	Polyol/polyester/ polyetherIsocyanate	Isocyanate is carcinogenic
Spray polyurethane foam (SPF)	Isocyanate (diphenylmethane diisocyanate)	Diisocyanates are severe bronchial irritants and asthmagens
Thermoplastic polyurethane foam (TPU)	Polyols and diisocyanates	TPU is made from a variety of highly hazardous intermediary chemicals, including formaldehyde (a known carcinogen) and phosgene (a highly lethal gas used as a poison gas in World War One, that, in turn, uses chlorinegas as an intermediary)
		In combustion, polyurethanes emit hydrogen cyanide and CO
Polyisocyanurate rigid foam (PIR or POLYISO)	Methylene diphenyl diisocyanate with a polyester derived polyol	PIR generally releases a higher level of toxic products in fires than other insulations
Expanded polystyrene foam (EPS) well known as Styrofoam	Closed cell foam made from Styrene	
Extruded polystyrene foam (XPS)		
Phenolic foam (PHF)	Phenolic resins made with a surface acting agent	May be made with formaldehyde
Composite insulation products: • insulated plasterboard incorporating the above materials • insulated panels incorporating the above materials • structural insulated panels (SIPs)		

Figure 6.1 Common foam insulation terminology
Source: Stec and Hull;[70] Brufma British Rigid Urethane Foam Manufacturers' Association[71]

serious complaints about health problems resulting from spray foam. 'Spray polyurethane foam insulation may cause health problems…respiratory and breathing problems, skin irritation and neurological issues'.[72,73] In addition to the use of isocyanates to create the foam, a catalyst to assist in curing may be an amine compound or lead naphthenate. Amine is derived from ammonia and can produce the fishy off-gassing smell that leads to complaints from occupants of buildings where the foam has been sprayed.

The manufacturers and installers claim that spray foam is safe and that the chemicals used off-gas over a very short time. However in the USA many hundreds of people have moved out of their houses claiming that spray foam made them ill and that odours lingered for months. It is easy to find hundreds of websites documenting these problems and lawyers' sites trawling for 'victims'. There are also claims that it is dangerous for installers and that there have been deaths of workers from spray foam bursting into flames.[74] In the UK there seems to be little concern about these problems even though advocates of spray foam warn of dangers. 'Spray foam insulation ideally needs to be installed by a professional, since it gives off dangerous fumes'.[75] It has even been suggested in a post by Mike George in Green Building Forum that polyurethane foam can be eaten: 'polyurethane foams which are inert and can actually be eaten without toxic effect but not recommended!'.[76]

There is considerable documentation about the dangers of sensitisation to isocyanates used in spray foam, leading to asthma problems. Once victims are sensitised then exposure to relatively low levels can trigger an attack. Most medical investigation of this problem has been concerned with occupational problems and the Health and Safety Executive issued a 252-page guidance document about the risks in 2005.

> After sensitisation, asthmatic attacks may occur at very low exposure levels, which may be much less than the relevant occupational exposure limit. The asthmatic condition may occur a few minutes after exposure, or more commonly several hours later, hence the characteristic nocturnal wakening. The severity of the asthmatic attacks which occur after re-exposure have been known to cause death in sensitised individuals.[77]

Shipping containers are hazardous

A particular dangerous use of spray foam insulation is in storage containers, some of which have been converted into housing for homeless people, such as in Brighton. Many seem to regard this as an environmentally friendly thing to do as it is recycling the containers. Many shipping containers are also used as offices. Given the airtight nature of steel containers, this means that there will be greater exposure to any off-gassing from insulation and other materials (such as treated flooring) in the containers.[78]

The British Urethane Foam Contractors Association (BUFCA) reassures us that spray polyurethane is safe on the basis that it is widely used, hardly a reassuring argument.

> Polyurethane is all around us in our everyday life, from the cars that we drive, furniture in our homes and work places, to the clothes that we wear. Whenever you get into a car, open a fridge, lie on a hospital bed or put on a pair of trainers, these are likely to contain polyurethane.[79]

While the industry claims that spray foam is not hazardous, it issues extensive guidance on how to dispose of the highly hazardous empty drums that the chemicals are transported in![80]

Despite the fact that many thousands of people claim to have been made ill by spray polyurethane, standards have not yet been adopted for levels of emissions from spray foam. 'Standardized methods are needed to assess the potential impacts of SPF insulation products on indoor air quality and to establish re-entry or re-occupancy times after product installation in a building and post-occupancy ventilation needs'.[81]

> Residential occupants re-entering SPF treated homes have complained about odor, severe asthma, and other respiratory symptoms. The severity of the health effects has in some cases caused homeowners to sell or abandon their homes. More information is needed on chemical emissions from SPF to better understand occupant exposures and their impacts on potential health effects.[82]

Risks from polyurethane in general

Apart from the hazardous flame retardants in polyurethane, the main constituents, isocyanates, are seen as a major health risk. Considerable research has been carried out into the dangers from isocyanates, which are regarded as asthmagens and possible carcinogens, but much of the literature focuses on occupational risks rather than to occupants in buildings.

> Isocyanates have a wide application in the industrial and domestic environment, their annual worldwide production being in the range of three million tons. The isocyanate group –N=C=O reacts with compounds containing active hydrogen atoms. This property makes diisocyanates useful for the production of polyurethane foam, coating material, and adhesive manufacture. The most commonly used diisocyanate is toluene diisocyanate (TDI), a mixture of 2,4 and 2,6 isomers (80:20). Other diisocyanates include hexamethylene diisocyanate (HDI), naphthalene diisocyanate (NDI), isophorone diisocyanate, and methylene diphenyl diisocyanate (MDI). TDI, NDI, and MDI prepolymers are also available, and they have been shown to give rise to occupational asthma. In spite of the recognised harmful health aspects of diisocyanates, they are widely used in most developed countries and therefore isocyanate induced asthma is one of the most studied forms of occupational asthma … The spectrum of lung diseases that can be induced by diisocyanates is broad, including asthma with latency, irritant induced asthma, and hypersensitivity pneumonitis.[83]

The inhalation of isocyanates has been associated with a range of complaints, including coughing, wheezing, chest discomfort, acute oedema and interstitial pulmonary fibrosis, as well as covert decrement of lung function.[84]

> Reported health effects in factories that produce polyurethane include: blurred vision, skin, eye and respiratory tract irritation, asthma, chest discomfort, etc. Some of the chemicals causing these symptoms outgas rather quickly, but others do not. Isocyanates are sensitizers. This means that they can sensitize a person and, once sensitized, that person will react to lower levels. Once these foams are cured, they no longer act as sensitizers.
>
> Polyurethane is flammable and must be separated from the living space by drywall or plaster. It burns rapidly and releases carbon monoxide, oxides of nitrogen, and hydrogen cyanide. Hydrogen cyanide is lethal (it's used in gas chambers) but so much carbon monoxide is released in a fire that it's of more concern. A group of firemen, who were exposed to isocyanates, reported numerous neurological symptoms such as: euphoria, headache, difficulty concentrating, poor memory, and confusion.[85]

While the polyurethane foam industry denies that there are any dangerous emissions from their products, and asserts that any off-gassing takes place at an early stage in manufacture and then dwindles to very low levels, chemicals used in these products are commonly found in IAQ tests. Scientific literature on off-gassing from synthetic insulations is not easy to find and there is an urgent need for more research on this topic. One area of concern is about off-gassing of toluene diisocynaate (TDI) from foam cushions and mattresses. Many people now sleep on mattresses made from TDI and thus are likely to breathe in significant levels of any chemicals emitted. A risk assessment published in *Regulatory Toxicology and Pharmacology* provides reassurance that this is not a health risk, however the authors of this paper worked for the chemical industry, Dow Chemicals Ltd, Hunstman, BASF and the International Isocyanate Institute. 'Although available data indicates TDI is not carcinogenic, a theoretical excess cancer risk was calculated. We conclude from this assessment that sleeping on a PU foam mattress does not pose TDI-related health risks to consumers'.[86] It is also misleading to state that TDI is not carcinogenic when there have been numerous studies suggesting that it is.[87,88]

> TDI is regulated under the Clean Air Act as a hazardous air pollutant and under RCRA and CERCLA as a hazardous waste. EPA will continue to work with its federal partners, the polyurethanes industry, and others to ensure improved labelling and product safety information for polyurethane products containing unreacted isocyanates, especially products targeted to consumers. EPA is also considering a green chemistry challenge to encourage the development of safer alternative chemicals.
>
> On May 12, 2010, Canada added TDI to its List of Toxic Substances following an assessment which determined that TDI is carcinogenic and affects the respiratory system. Environment Canada published a proposed P2 Pollution Prevention Planning Notice for the polyurethane and other foam sector. The International Agency for Research on Cancer (IARC) has classified TDI as possibly carcinogenic to humans.[89]

A frequently cited study by Collins called the 'GIL report', claiming that TDIs are not a health risk and that off-gassing is minimal, appears to have come from a marketing company called Gilbert International working for the chemical industry. Also the main reference book on TDI states that polyurethanes pose no threat to air quality but was funded by the International Isocyanates Institute (III) and was edited by three members of Gilbert International based in Manchester, also the head office of the III, though neither of these could be traced.[90]

> The GIL Report concludes that the levels of TDI in fully cured foams are below the limits of detection (less than 2 ppb). While some freshly manufactured foams contain trace amounts of TDI, these levels appear to quickly fall to levels below the limits of detection of the test methods when the foam is fully cured – within several hours of production.[91]

Phenolic foam

PHF is a thermoset cellular polymer made from phenolic resins, reacted with a catalyst, hardening agent and a blowing agent to foam the product. The catalyst is normally an acid, toluene and/or xylene sulfonic acid. The chemicals used as catalysts are hazardous. PHF can have an open or closed cell structure and is claimed to have better fire and smoke properties than other foam insulations such as polyurethane.[92] However, fire testing has shown that all foam insulations present fire risks. Fire testing has conventionally been done in well-ventilated conditions and this has not provided an accurate assessment of toxic emissions.[93]

Expanded and extruded polystyrene

EPS is a thermoplastic polymer. Manufacture is normally by polymerisation of styrene with pentane to produce long polymer chains incorporating pentane in solution. The granules will expand under the influence of steam (prefoaming) to produce beads that can be further processed to produce blocks and boards or shaped mouldings.

Styrene is classified as toxic for reproduction and organ toxic. The US NTP described styrene as 'reasonably anticipated to be a human carcinogen'.[94] Styrene is vinyl benzene. Benzene is classified as a carcinogen. Styrene-induced neurotoxicity has been reported in workers since the 1970s with both short- and long-term occupational exposures to styrene possibly resulting in neurological effects. Styrene is also classified as a xenoestrogen, an endocrine-disrupting compound. These are toxic chemicals that mimic hormones and can seriously disrupt the endocrine system.[95] There is also evidence that styrenes can leach out of polystyrene and foam food packaging and into food that is consumed.[96]

Despite these concerns, it is commonly assumed that polystyrene is less of a health hazard than polyurethane. This is because it is assumed that off-gassing from styrenes is claimed not to linger. However insufficient research has been done to assess the risks from styrenes used in homes. Building Green recommends avoiding polystyrene, not so much because of the problem with styrenes but because of the problems of flame retardants used with polystyrene.[97]

Avoid using foam insulations where possible

The manufacture and risks from foam insulations, particularly the sprayed materials, is a murky and worrying area of science, with little convincing independent research except to confirm risks and hazards. There is little reassurance about the safety of these materials in terms of their impact on IAQ and in fires.

Notes

1 Email from Sarah Wilde Buildings Societies Association, 3 March 2016
2 UK damp & decay control www.ukdamp.co.uk [viewed 20.2.16]
3 UK damp & decay control op cit.
4 Toxic treatments, London Hazards Centre www.lhc.org.uk/wp-content/uploads//Resources/Books/lhc_toxic_treatments_wood_preservative_hazards_206_pgs_january_1989.pdf [viewed 20.2.16]
5 WWF (2003) *Contamination: The Results of WWF's Biomonitoring Survey*, November, WWF
6 Litmans, B. and Miller, J. (2004) *Silent Spring Revisited*. Centre for Biological Diversity
7 www.hse.gov.uk/biocides/copr/creosote.htm [viewed 23.3.16]
8 www.archtimberprotection.com/wp-content/uploads/2015/05/Tanalith-Family-Treated-Timber-User-Guide.pdf [viewed 23.3.16]
9 Smith, C. (2004) Good Enough To Eat: Copper-based preservative also has risks www.seattlepi.com/lifestyle/homegarden/article/Good-Enough-To-Eat-Copper-based-preservative-1142205.php [viewed 20.2.16]
10 Extension Toxicology Network, Pentachlorophenol http://pmep.cce.cornell.edu/profiles/extoxnet/metiram-propoxur/pentachlorophenol-ext.html [viewed 20.2.16])
11 Stockholm convention on persistent organic pollutants www.pops.int/documents/convtext/convtext_en.pdf [viewed 20.2.16]
12 OSHA Hazard Communication Standard (29 CFR 1910.1200) www.osha.gov/dsg/hazcom/standards.html [viewed 27.2.16]
13 Badreshia, S. and Marks, J. (2002). Iodopropynyl Butylcarbamate'. *American Journal of Contact Dermatitis* 13 (2), 77–79
14 Safety of woodworm treatments www.woodworm-expert-advice-forum.org.uk/safety.htm [viewed 20.2.16]

15 BWPDA – British Wood Preserving and Damp-proofing Association www.property-care.org/us/bwpda/ [viewed 20.2.16]
16 ELINCS, REACH Online www.reachonline.eu/REACH/EN/REACH_EN/kw-elincs.html [viewed 27.2.16]
17 Least toxic timber treatment www.pan-uk.org/pestnews/Homepest/timber.htm [viewed 20.2.16]
18 Timber treatment, Centre for Alternative Technology http://info.cat.org.uk/sites/default/files/documents/TimberTreatment.pdf [viewed 20.2.16]
19 Toxic wood, Health and Safety Executive www.hse.gov.uk/pubns/wis30.pdf [viewed 20.2.16]
20 Hazardous substances www.hse.gov.uk/woodworking/hazard.htm [viewed 20.2.16]
21 Toxicological riddles: the case boric acid, Building Green http://greenspec.buildinggreen.com/blogs/toxicological-riddles-case-boric-acid#sthash.iOo2XqWW.dpuf [viewed 20.2.16]
22 Blown-in, cellulose-based insulation (2016) *Natureplus* www.natureplus.org/fileadmin/user_upload/pdf/cert-criterias/GL0107.pdf (viewed [viewed 27.2.16]
23 Rio Tinto Minerals www.riotintominerals.com [viewed 20.2.16]
24 Olympic medal mining firm Rio Tinto faces air pollution lawsuit in US, *The Guardian*, 24 September 2013 www.theguardian.com/business/2013/sep/24/olympic-medal-mining-firm-rio-tinto-air-pollution-lawsuit-us [viewed 20.2.16]
25 From rock to roof, Eden Project www.edenproject.com/sites/default/files/rock-to-roof-journey-of-copper.pdf [viewed 27.2.16]
26 WWF (2003) op cit.
27 Healthy buildings, Green Science Policy Institute greensciencepolicy.org/topics/healthy-buildings [viewed 20.2.16]
28 Harrad, S. (2009) *Persistent Organic Pollutants December*, London: Wiley-Blackwell
29 Mishamandani, S. (2014) Lecture highlights flame-retardants, National Institute of Environmental Health Sciences, Environmental Factor
30 Patisaul, H.B., Roberts, S.C., Mabrey, N., McCaffrey, K.A., Gear, R.B., Braun, J., Belcher, S.M. and Stapleton, H.M. (2013) Accumulation and endocrine disrupting effects of the flame retardant mixture Firemaster® 550 in rats: an exploratory assessment www.ncbi.nlm.nih.gov/pubmed/23139171 [viewed 20.2.16]
31 Pillai, H.K. *et al.* (2014) Ligand binding and activation of PPARy by Firemaster ® 550 Effects on Adipogenesis and Ostwogwnwsus in Vitro. *Environmental Health Perspectives* 122 (11) November
32 Bergman, A., Heindel, J.J., Jobling, S., Kidd, K.A. and Zoeller, R.T. (2012) *State of the Science of Endocrine Disrupting Chemicals*, WHO
33 Babrauskas, V., Lucas, D., Eisenberg, D., Singla, V., Dedeo, M. and Blum, A. (2012) Flame retardants in building insulation: a case for re-evaluating building codes, *Building Research & Information* 40 (6), 738–755 http://dx.doi.org/10.1080/09613218.2012.744533 [viewed 20.2.16]
34 Babrauskas, V. *et al.* (2012) op cit.
35 Reducing Flame Retardants in Building Insulation to Protect Public Health (2015) Alpha, Policy Number: 20156 www.apha.org/policies-and-advocacy/public-health-policy-statements/policy-database/2016/01/05/18/39/reducing-flame-retardants-in-building-insulation-to-protect-public-health [viewed 20.2.16]
36 Grossmn, E. (2011) Are Flame Retardants Safe? Growing Evidence Says 'No', *Environment 360* http://e360.yale.edu/feature/pbdes_are_flame_retardants_safe_growing_evidence_says_no/2446/ [viewed 20.2.16]
37 www.adhesives.org/adhesives-sealants/adhesives-sealants-overview/health-safety [viewed 23.3.16]
38 Priha, E., Hellman, S. and Sorvari J. (2005) Contamination from polysulphide sealants in residential areas-exposure and risk assessment. *Chemosphere* April, 59 (4), 537–543
39 www3.epa.gov/epawaste/hazard/tsd/pcbs/pubs/caulk/pdf/pcb_bdg_mat_qa.pdf [viewed 23.3.16]
40 www.ehow.co.uk/list_7565387_effects-silicone-caulking-hands.html [viewed 23.3.16]

41 Major countries in worldwide asbestos mine production from 2010 to 2015 (in metric tons), The statistics portal www.statista.com/statistics/264923/world-mine-production-of-asbestos/ [viewed 20.2.16]
42 www.asbestos.com/mesothelioma/worldwide.php#worldwide [viewed 13.3.16]
43 Asbestos statistics show 'devastating legacy', *Hazards magazine* 132 (2015), 26
44 Fleming, N. (2014) Killer dust: Why is asbestos still killing people www.newstatesman.com/sci-tech/2014/03/killer-dust-why-asbestos-still-killing-people [viewed 20.2.16]
45 The Asbestos Crisis: Why Britain needs a new law (2015) All Party Parliamentary Group on Occupational Safety and Health www.voicetheunion.org.uk/index.cfm?cid=1537 [viewed 23.3.16]
46 *Hazards Magazine* (2015) 132, 27
47 Teacher died of asbestos poisoning (2016) *Morning Star* www.morningstaronline.co.uk/a-ab70-Teacher-died-of-asbestos-poisoning#.Vqzt6DmQSN9 [viewed 20.2.16]
48 Asbestos cement, Asbestos Information Centre www.aic.org.uk/asbestos-cement/ [viewed 20.2.16]
49 Woolley, T. (2013) *Low Impact Building*, London: Wiley, pp. 4–5
50 *Report on Carcinogens*, Thirteenth Edition https://ntp.niehs.nih.gov/ntp/roc/content/profiles/glasswoolfibers.pdf [viewed 20.2.16]
51 Montague, P. (1995) Fibreglass: a carcinogen that's everywhere, Greenleft www.greenleft.org.au/node/10730#sthash.uvaXQrfn.dpuf [viewed 20.2.16]
52 Whistleblower Exposes Shocking Evidence About A Carcinogen That is Absolutely Everywhere (2012) Prevent Disease http://preventdisease.com/news/12/040312_Whistleblower-Exposes-Shocking-Evidence-About-A-Carcinogen-That-is-Absolutely-Everywhere.shtml [viewed 20.2.16]
53 Is the Fiberglass in Your Attic or Walls Causing Cancer? (2012) Mercola http://articles.mercola.com/sites/articles/archive/2012/04/30/fiberglass-causes-cancer.aspx [viewed 20.2.16]
54 Lent, T. (2009) *Formaldehyde emissions from fiberglass insulation with phenol formaldehyde binde*r, Healthy Building Network http://healthybuilding.net/uploads/files/formaldehyde-emissions-from-fiberglass-insulation-with-phenol-formaldehyde-binder.pdf [viewed 20.2.16]
55 Axten, C.W. (2003) *Formaldehyde and Total Volatile Organic Compound Emissions from Thermal Insulation Products*, Presentation to the Formaldehyde Product Stewardship Committee. http://healthybuilding.net/uploads/files/formaldehyde-emissions-from-fiberglass-insulation-with-phenol-formaldehyde-binder.pdf [viewed 24.3.16]
56 http://thehealthcoach1.com/?p=1995 [viewed 24.3.16]
57 Vallette, J. (2015) Residential Fiberglass Insulation Transformed: Formaldehyde is No More, *Healthy Building News* www.healthybuilding.net/news/2015/10/30/residential-fiberglass-insulation-transformed-formaldehyde-is-no-more#sthash.Bxo6Ak8z.dpuf [viewed 20.2.16]
58 Brown Insulation, Knauf Insulation www.knaufinsulation.co.uk/products/brown-insulation [viewed 20.2.16]
59 Owens Corning (2012) Just the Facts Formaldehyde and Insulation www2.owenscorning.com/literature/pdfs/57470.pdf
60 Owens Corning (2012) op cit.
61 Envirotest www.envirotest.eu ([viewed 20.2.16]
62 Woolley (2013) op cit.
63 Zero Carbon Hub (2013) Closing the gap between design and as-built performance www.zerocarbonhub.org/sites/default/files/resources/reports/Closing_the_Gap_Bewteen_Design_and_As-Built_Performance_Interim_Report.pdf [viewed 20.2.16]
64 Pentane, Scottish Pollutant Release Inventory, SEPA http://apps.sepa.org.uk/spripa/Pages/SubstanceInformation.aspx?pid=79 (viewed 27.02.16)
65 Plastinum Foam – Foaming with CO_2, Linde Plastic www.lindeplastics.com/foaming-co.php [viewed 20.2.16]

66 Extruded Polystyrene Foam Board Insulation, Honeywell Solstice Gas Blowing Agent www.honeywell-blowingagents.com/?document=solstice-gba-in-xps-insulation&download=1 [viewed 20.2.16]

67 Lakanal House tower block fire: deaths could have been prevented, *The Guardian* www.guardian.co.uk/uk/2013/mar/28/lakanal-house-fire-deaths-prevented [viewed 20.2.16]; Council and contractor clash at fire death inquest (2013) *Inside Housing* www.insidehousing.co.uk/legal/council-and-contractor-clash-at-fire-death-inquest/6525972.article [viewed 20.2.16]

68 Council and contractor clash at fire death inquest (2013) *Inside Housing* www.insidehousing.co.uk/legal/council-and-contractor-clash-at-fire-death-inquest/6525972.article [viewed 20.2.16]

69 Dubai blaze raises questions over Gulf skyscraper design, Reuters http://uk.reuters.com/article/emirates-dubai-fire-idUKKBN0UG0H720160102 [viewed 20.2.16]

70 Stec, A. and Hull, R. (2011) Assessment of the fire toxicity of building insulation materials, *Energy and Buildings*, 43 (2-3), 498–506

71 Brufma British rigid urethane foam manufacturers' association http://brufma.co.uk [viewed 20.2.16]

72 Spray polyurethane foam can cause serious health problems www.yourlawyer.com/topics/overview/spray-foam-insulation-risks-lawsuits [viewed 20.2.16]; Does Spray Foam Insulation Off-Gas Poisonous Fumes?, Green Building Advisor www.greenbuildingadvisor.com/blogs/dept/qa-spotlight/does-spray-foam-insulation-gas-poisonous-fumes [viewed 20.2.16]

73 Does Spray Foam Insulation Off-Gas Poisonous Fumes? op cit.

74 www.greenbuildingadvisor.com/blogs/dept/green-building-news/three-massachusetts-home-fires-linked-spray-foam-installation [viewed 20.2.16]

75 Spray Foam Insulation FAQ, The Green Age www.thegreenage.co.uk/spray-foam-insulation-faq/ [viewed 20.2.16]

76 Green Building Forum www.greenbuildingforum.co.uk/newforum/comments.php?DiscussionID=2847 [viewed 20.2.16]

77 Institute of Occupational Medicine for the Health and Safety Executive (2005) An occupational hygiene assessment of the use and control of isocyanates in the UK, London: Institute of Occupational Medicine for the Health and Safety Executive

78 The charity that houses homeless people in shipping containers (2013) *The Guardian* www.theguardian.com/housing-network/2013/oct/24/brighton-housing-homeless-shipping-containers [viewed 27.2.16]

79 BUFCA, Health and Safety Information www.bufca.co.uk/wp-content/uploads/2015/02/Health-Safetypdf.pdf [viewed 27.2.16]

80 Disposal of Used Spray Polyurethane Foam Drums, Safety First https://spraypolyurethane.org/disposalSPFdrums [viewed 20.2.16]

81 Symposium on Developing Consensus Standards for Measuring Chemical Emissions from Spray Polyurethane Foam (SPF) Insulation, Sponsored by ASTM Committee D22 on air Quality April 30–May 1, 2015 Marriott Anaheim

82 Poppendieck, D.G., Nabinger, S.J., Schlegel, M.P. and Persily, A.K., (2014) Long Term Emissions from Spray Polyurethane Foam Insulation, 13th International Conference on IAQ and Climate Conference 2014, July 7–12, Hong Kong

83 Mapp, C.E. (2001) Agents, old and new, causing occupational asthma, *Occupational & Environmental Medicine*, 58, 354, doi:10.1136/oem.58.5.354 [viewed 23.3.16]

84 UK Health and Safety Directorate (2003) Methods for the Determination of Hazardous Substances Organic Isocyanates in air, January, UK Health and Safety Directorate

85 Polyurethane and polyisocyanurate, Healthy House Institute www.healthyhouseinstitute.com/hhip-505-Polyurethane-and-polyisocyanurate#sthash.i1um0PXQ.dpuf [viewed 20.2.16]

86 Arnold, S.M., Collins, M.A., Graham, C., Jolly, A.T., Parod, R.J., Poole, A., Schupp, T., Shiotsuka, R.N. and Woolhiser, M.R. (2012) Risk assessment for consumer exposure to toluene

diisocyanate (TDI) derived from polyurethane flexible foam, *Regul Toxicol Pharmacol* 64(3), 504–515

87 Bolognesi, C., Baur, X., Marczynski, B., Norppa, H., Sepai, O. and Sabbioni, G. (2001) Carcinogenic risk of toluene diisocyanate and 4,4'-methylenediphenyl diisocyanate: epidemiological and experimental evidence, *Crit Rev Toxicol* 31(6), 737–772

88 The International Agency for Research on Cancer (IARC) has classified 2,4-toluene diisocyanate as a Group 2B, possible human carcinogen. www3.epa.gov/airtoxics/hlthef/toluene2.html [viewed 20.2.16]

89 US Environmental Protection Agency (2011) Toluene Diisocyanate (TDI) and Related Compounds Action Plan, EPA

90 Allport, A.C., Gilbert, D.C. and Outterside, S.M. (2003) *MDI and TDI Safety Health and the Environment: A Source Book*, Wiley

91 Collins, M.A. (2000) Diisocyanates and Indoor Air, GIL Report 2000/C, Gilbert International Limited, Manchester, UK www.chemicaldesigncorp.net/PDF_Files/Off_Gasing_Reference.pdf [viewed 20.2.16]

92 Tingley, D.D., Hathaway, A., Davison, B. and Allwood, D. (2014) The environmental impact of phenolic foam insulation boards, *Proceedings of the Institution of Civil Engineers* http://dx.doi.org/10.1680/coma.14.00022 Paper 1400022 [viewed 27.2.16]

93 Stec, A., Hull, Richard T. (2011) Assessment of the fire toxicity of building insulation materials. *Energy and Buildings* 43 (2–3), 498–506 ISSN 03787788 http://dx.doi.org/10.1016/j.enbuild.2010.10.015 [viewed 27.2.16]

94 Styrene (2011) Report on Carcinogens, Thirteenth Edition, National Toxicology Program, Department of Health and Human Services

95 Toxicological profile for styrene US Department of Health and Human Services, Public Health Service Agency for Toxic Substances and Disease Registry November 2010

96 Styrene Among Top Suspected Carcinogens – American Cancer Society, Eco Panels www.eco-panels.com/blog/63-eps-as-a-carcinogen.html [viewed 20.2.16]

97 Wilson, A. (2009) Avoid Polystyrene Insulation www2.buildinggreen.com/blogs/avoid-polystyrene-insulation [viewed 20.2.16]

7 Mould, damp, fuel poverty and breathability

A crucial aspect of bad IAQ is a result not of toxic chemicals from building materials or other synthetic substances introduced into the home, but from an entirely natural source. Damp and mould are a major threat to healthy indoor environments and must be managed as effectively as possible as these can be one of the main causes of respiratory illnesses and discomfort. This chapter discusses the nature and causes of damp problems but also offers alternative solutions as conventional solutions to damp and mould can often be unsuccessful and introduce even more toxic chemicals into buildings. Innovative solutions using natural and moisture buffering materials are worthy of further development and thus building materials are critical when considering damp and mould.

A survey carried out in 2014 for the UK Energy Saving Trust found that 39 per cent of respondents claimed to have condensation problems in their home and 28 per cent had mould problems. In reality the incidence of damp, condensation and mould is likely to be even higher as many people are unaware of it or find it embarrassing to admit to the problem as mould growth is associated with having a dirty home.[1]

Fuel poverty

Fuel poverty is a major problem in the UK and is closely associated with mould and damp. Many people cannot afford to heat their homes in what is claimed to be the sixth richest country in the world. The official definition of fuel poverty is households that have to spend more than 10 per cent of their annual income on heating. However this is not just a case of financial poverty and fuel costs but also a result of badly built, badly designed and poorly insulated houses. Many people classified as living in fuel poverty live, not in rotting 19th-century hovels, but modern recently constructed dwellings, including those that are classified as being energy efficient. In England over 10 per cent of the population (2.35 million people) were officially in fuel poverty in 2013. In Scotland in 2010, 658,000 people were in fuel poverty and in Northern Ireland it was 42 per cent, nearly half the population![2] There is a massive industry of fuel poverty experts and millions of pounds are spent every year on statisticians, fuel poverty ‘experts’, consultants and a range of charities and non-government organisations, discussing how to define fuel poverty, the latest statistics and how to solve the problem. However the problem remains intractable and little is being done to relieve the suffering of people who have to put up with cold houses and older people who will get sick or even die because they cannot keep their houses warm. Despite warmer winters, winter ‘excess’ deaths reached 44,000 in 2014.

It was impossible to find a single insightful critique of the current approach to fuel poverty, despite the concern of politicians and those who have to deal with the problem on a daily basis. A range of programmes to improve or replace poor quality housing, to grant aid replacement boilers and provide insulation grants have been a dismal failure, poorly targeted, ineffective and poorly directed. This has left many houses in conditions that exacerbate damp and mould and worsen respiratory conditions. Underlying this is a lack of expertise and knowledge, where it is assumed that installing loft insulation, new boilers and solar panels will deal with the problem.

Explaining the failures of the fuel poverty industry will be the subject of another book, but what is stunningly clear from an extensive review of the literature is that many people involved in the fuel poverty industry do not know enough about building technology and how to keep people warm. They also fail to address the serious issue of damp and mould and the massive burden this places on hospitals, doctors and the UK NHS. Damp and mould are inextricably linked with badly insulated and ventilated houses and so fuel poverty is an IAQ issue.

Mould

Mould spores are always present, and while invisible, are floating around in the air or contained in dust, often in huge concentrations both indoors and outdoors, looking for a home to settle and breed! Conditions that support mould growth are significantly affected by building construction methods and materials.

Much has been written about this problem and the related issue of dust mite allergies, however very little of the literature makes a clear link to building materials and how different forms of building construction can exacerbate or reduce the impact of mould and damp. Howieson's excellent book, *Housing and Asthma*[3] explains the link between dampness in houses, mould and house dust mites, which is one of the main causes of allergic reactions, leading to asthma. He, unlike many others, does refer to the importance of thermal mass, natural insulation and hygroscopic materials, to reduce the incidence of problems, and this is explored more fully in this chapter.

In addition to asthma, a range of respiratory problems can result from mould and damp. These include:

- nasal and sinus congestion; runny nose
- eye irritation; itchy, red, watery eyes
- respiratory problems, such as wheezing and difficulty breathing, chest tightness
- cough
- throat irritation
- skin irritation, such as a rash
- headache
- persistent sneezing

Even more serious problems result from 'toxic mould'. Toxic mould is produced from mycotoxins, such as *Stachybotrys chartarum*.

Moulds can grow on almost any substance when moisture is present. They reproduce by distributing spores, which are carried by air currents. When spores land on a moist surface, suitable for life, they begin to grow. Mould is normally found indoors in all buildings, but usually at levels that do not affect most healthy individuals. Because common building materials are capable of sustaining mould growth and mould spores are endemic, mould growth in an indoor environment is typically related to the presence of water or moisture and may be caused, for example, by incomplete drying of flooring materials (such as concrete). Flooding, leaky roofs, building maintenance or indoor-plumbing problems can lead to interior mould growth. Water vapour commonly condenses on surfaces cooler than the surrounding moisture-laden air, enabling mould to flourish. Moisture vapour passes through walls and ceilings, typically condensing interstitially during the winter in climates with a long heating season.

Significant mould growth requires moisture and food sources and a substrate capable of sustaining growth. Common building materials, such as plywood, plasterboard (drywall), carpets and carpet underlay provide food for mould. In carpet, invisible dust and cellulose are food sources. After water damage to a building, mould grows in walls and then becomes dormant until subsequent high humidity conditions reactivate the mould. Mycotoxin levels are higher in buildings that have had a flooding or leakage incident. This is likely to become a more significant problem with global warming and greater incidence of extreme weather and flooding.

The first step in solving an indoor mould problem is to remove the moisture source, as new mould will begin to grow on moist, porous surfaces within 24 to 48 hours. There are a number of ways to prevent mould growth. Some cleaning companies specialise in fabric restoration, removing mould (and mould spores) from clothing to eliminate odour and prevent further damage. The normal way to clean mould is to use detergents in an effort to physically remove it. Many commercially available detergents include antifungal agents but care should be taken to ensure that such products have a low toxicity. Detergents and bleach are not always effective at entirely removing mould.

Significant mould growth may require professional mould remediation to remove the affected building materials and eradicate the source of excess moisture but in extreme cases demolition may be the only answer! Workers carrying out remediation works also need to have high levels of protection from the mould spores.

Problems with insulation and retrofit solutions

Some materials are more receptive than others to mould growth and moist surfaces can result from bad design such as 'cold bridging'. If a building is badly designed and poorly ventilated, condensation can occur and water can be seen forming on colder surfaces such as windows, window frames and walls and ceilings. Thermal bridging (which is a more accurate term than cold bridging), where insulation is not continuous, is a major cause of damp and mould growth and can often be seen around lintels and in areas where the insulation has been bridged. Mould often develops in corners and behind furniture where there is less air movement. Mould will also result from interstitial condensation, in other words within building fabric where it may not be apparent to the eye.

A study carried out by BRE for the Welsh Government included the surprising information that cavity wall insulation (CWI) had to be removed from 280 houses in Wales because of dampness problems and a further 900 houses required remedial action.[4] The BRE study had not been published and details were only released following a freedom of information request from the Cavity Insulation Victims Alliance.[5,6] It has been suggested that CWI may exacerbate dampness in walls and may not be as effective as is claimed. So serious is this problem of dampness caused by badly installed cavity insulation that it has been claimed that 3 million houses in England may have been affected and this was debated in the UK House of Parliament in February 2014.[7]

Lawyers are now acting on behalf of the so-called victims of CWI and claims are being made against installers.[8] Several installers have gone out of business but there is a guarantee scheme designed to protect customers.[9] However a major study of CWI has drawn attention to dampness problems.[10] Despite this, UK government bodies and agencies such as the Energy Saving Trust (EST) continue to advocate CWI.[11] Questions exist about the EST and how it arrives at its policy advice. On its website it lists nine 'experts' but these are communication, marketing, policy and advisory people. Not one mentions insulation or retrofitting in their list of skills and experience. The non-executive directors of EST include a director of Rentokil Initial, an IT person from Barclays Bank, a chemical engineer and a sun tan and vitamin marketing executive from Boots and others with little apparent connection to energy policy. The EST appears to have nearly £16 million in its coffers and could do much more to engage with fuel poverty, damp and mould problems.[12]

Another serious problem is with solid walled buildings where there is little or no insulation. A recent study by the UK government's chief construction advisor Peter Hansford, refers to problems with solid wall insulation (SWI) measures.[13] This report, made to the Green Construction Board, makes a case for improving health and well being by treating properties with SWI, but highlights the lack of performance data from properties that have been retrofitted with SWI and that proper design of retrofit solutions is 'frequently absent entirely'. Because of the lack of design and coordination in retrofit projects there could be many problems.

> there is an unacceptable risk of unintended consequences, which might include damp, mould, poor air quality and poor building performance. Depending on the extent of the problems, this may cause health issues for occupants together with the possible failure to achieve the predicted improvement in thermal performance required to deliver lower energy bills and reduced greenhouse gas emissions. At the extreme, for some property archetypes, structural damage could be incurred.[14]

One unintended consequence of poor insulation and retrofit solutions is damp and mould problems and there are many buildings that have some mould growth, even though it may not always be visible. Even apparently clean and well-kept houses may suffer from mould and once mould is established, it can grow rapidly. Mould can be found in brand new buildings, as many buildings do not dry out properly before decoration and occupation. Mould may come with the building materials such as concrete blocks that have been out in the rain for months or even years. Concrete floors that have not dried out fully before coverings are installed may be another source of moisture and mould. Floods, leaks, plumbing problems and poor maintenance will lead to dampness and mould.

Even where buildings are well insulated or have been retrofitted, damp and mould can occur. Indeed high levels of insulation from poorly designed interventions can lead to greater levels of dampness and associated health problems, as indicated by Hansford above. Pressure to insulate and retrofit existing buildings to save energy has led to many inappropriate insulation measures that can result in dampness and mould growth. 'New Government schemes which promote the use of insulation, and make us install loft and cavity wall insulation are causing huge problems with condensation'.[15]

In particular, interstitial condensation can be caused by dry-lining walls with synthetic insulation materials and plastic membranes where moisture becomes trapped in the space between the insulation and external walls. These problems can be avoided or the risk substantially reduced through the use of vapour permeable materials, as will be discussed below, or by providing adequate ventilation to voids.

A study carried out by Exeter University of houses in Cornwall that had been made more energy efficient, reported that this had led to a small increase in asthma, with an increase in a mouldy musty odour in the dwellings that had been retrofitted.

> Higher energy efficient homes are associated with increased risk of doctor diagnosed asthma in a UK subpopulation. A unit increase in household Standard Assessment Procedure (SAP) rating was associated with a 2% increased risk of current asthma, with the greatest risk in homes with SAP N71. We assessed exposure to mould and found that the presence of a mouldy/musty odour was associated with a two-fold increased risk of asthma (OR 2.2 95%; CI 1.3–3.8).[16]

There is a general assumption in literature dealing with fuel poverty and poor housing conditions, that adding insulation will make houses warmer and this will lead to better health and reduced admissions to hospital. This is a very naive assumption and not always born out by the evidence. The idea that adding any sort of insulation to houses is the answer to health problems largely comes from social policy academics, who have little knowledge of building technology. One of the leading figures in the fuel poverty industry, and an advisor to government on energy saving policies was Professor Brenda Boardman. Boardman's highly influential book *Fixing Fuel Poverty* barely discusses remedial measures to houses or methods of insulating them and advocates measures such as combined heat and power (CHP) as a solution to fuel poverty! For instance Boardman suggests that installing central heating in homes will reduce childhood asthma without making the connection that the fuel poor cannot afford to run central heating![17]

Boardman and many others working in fuel poverty come from a social and economic policy background and thus their understanding of technical issues may be limited. They tend to rely far too heavily on quoting Standard Assessment Procedure (SAP)[18] figures as a measure of energy efficiency, whereas SAP, according to Scott Kelly has been shown to 'be a poor predictor of dwelling level energy demand'. Kelly goes on to argue that SAP is a poor measure of economic efficiency of a building.[19]

The fuel poverty 'experts', in assuming that loft insulation and central heating will improve the health of householders, fail to appreciate the complexity of issues about buildings and health as addressed in this book. The measures they advocate may be doing more harm than good by increasing VOC emissions, damp and mould growth, etc.

In a more recent report reviewing progress on tackling fuel poverty, Hills documents government funding cuts to insulation schemes and points out that the problem of fuel poverty is likely to be 'substantially worse', but entirely ignores the importance of finding well-designed and appropriate retrofit measures.[20]

Grants under fuel poverty schemes have been available to increase loft insulation, for instance. This work has been done by installation companies, many of which may have insufficient training and expertise. This can lead to problems, such as when additional loft insulation is installed in cold roof spaces. When insulation was previously poor, heat would leak from the rooms below, keeping the roof space warm enough to minimise condensation problems. However with the installation of extra insulation (often poorly installed, blocking up crucial roof vents), condensation can form on the cold roof finishes and so-called roof 'breather' membranes/roofing felt lead to water dripping onto the insulation. Once synthetic insulations such as mineral fibre/glass fibre become wet, they can be less effective as insulants.

There is a great deal of confusing and contradictory advice available to the public about how to tackle insulation, condensation and dampness. Most advice focuses on installing vapour control layers and mechanical ventilation, rather than using materials that can cope with moisture. 'Why is there condensation in my loft? Jeff Howell tackles insulation and condensation in the attic, and explains how to fit an extractor fan on a very thick wall'.[21] The scale of this problem is likely to increase, particularly as measures to insulate houses rarely take account of unintended damp problems, as can be seen from the following examples. 'A tenant secured £11,200.46 damages against a housing association for damp and disrepair'.[22] 'Loft and cavity wall insulation causes damp'.[23] 'Reports of damp soar in social housing as residents avoid turning on heating. Condensation dampness – regarded as major public health risk – is said to be increasing, with experts blaming rising energy bills'.[24] 'Damp has ruined everything, says housing association tenant'.[25] 'Sutton mum-of-three battles with housing association over mould'.[26] 'Borehamwood mother's mould-ridden house is "depressing to live in"'.[27]

Affinity Sutton, a large UK housing association, carried out a very important and brave study, *Future-Fit – Living with Retrofit*, published in 2013, about a programme in which 102 homes were retrofitted. After insulation was added, 73% of their tenants said that the homes were warmer, but 78 tenants reported problems of which 17% were directly related to damp, mould or condensation. Many also complained of ventilation problems and the heat recovery ventilation systems were 'too noisy'. The most worrying result was that the SAP model used in Future-Fit *over predicted* savings by an average of 77 per cent. 'Condensation can increase after retrofit works, even … (including) works to install cavity and loft insulation'.[28]

The finding that SAP over predicts energy performance has become known as the 'performance gap', which is reflected in many other projects where retrofit solutions have failed to be very effective and have led to unintended consequences of damp, etc.[29] Similar results were found in the Retrofit for the Future programme, massively funded by the TSB (now Innovate UK) where only four projects out of 115 came anywhere near meeting the predicted energy saving targets and the majority used 50 per cent more energy than had been predicted.[30]

Organisations such as the Zero Carbon Hub and academics in the field of construction management[31] focus almost entirely on poor construction methods and training as the cause of the problems referred to above. However they largely ignore the materials and technical solutions as being at fault. Blaming the builders is easy, but why not blame the synthetic plastic materials that make up 95 per cent of normal building and retrofit practice? The intention to achieve greater airtightness in buildings is to make them more energy efficient but this too can have unintended consequences in terms of health.

Moves towards increased air tightness could have a significant effect on humidity in buildings, particularly housing. It has been suggested that increased airtightness will make houses healthier as they are warmer and draft free, however some argue that this can be dangerous.

> I also believe that air-tightness can be taken to an unnecessary and potentially dangerous extreme. With Passive House standards becoming the ideal for some designers, it is valuable to ask, 'How tight is tight enough?' The respected building scientist, John Straube, who has had the opportunity to test and observe thousands of houses in Canada's cold climates, has offered an answer that agrees with my own experience. Homes with greater than 3 ACH50 (air changes per hour pressurized to 50 pascals with a blower door) tend to have a risk of interstitial condensation; those with greater than 5 or 6 ACH50 tend also to be too dry inside.[32]

Avoiding snake oil solutions

Householders and maintenance managers will respond to mould problems by cleaning. Usually bleach is used, adding yet another indoor air pollutant to the indoor air mix. However cleaning with bleach may only be a short-term solution with mould growth returning after some time. Mould spores take root in conventional materials such as cement and gypsum plasters. Painting over the problem with acrylic paints only makes matters worse.

At a seminar on healthy buildings in Northern Ireland, environmental health officers admitted that they had told landlords with damp and mould problems to stick thin polystyrene over problem areas as this would 'insulate' the walls.

The Labour Housing Minster Yvette Cooper in 2007 attracted ridicule by describing this solution as 'magic wallpaper':

> We need to develop radical environmental technologies that solve problems like that. The way I describe it is that you need 'magic wallpaper'. The officials think I'm stupid when I say that, but it is what you need: well-insulating, inexpensive stuff that you can put on to solid walls easily and that doesn't add an extra couple of inches to the walls.[33]

The general public are always vulnerable to 'snake-oil' solutions to problems and often get bad advice, even from a government minister. There have been numerous scams foisted on the public such as magic insulating paint using 'nano technology' and anti-condensation paint.[34] Dealing with dampness and mould through the use of breathable and hygroscopic materials is discussed in Chapter 9.

Moisture, mould and health problems

Moisture vapour is ever present and can also pass through building materials or if not vapour permeable, it can be trapped between layers. Places with high background humidity, often near the sea, can have the worst damp and mould problems, so cities like Glasgow, Belfast and Liverpool, with serious problems of fuel poverty, also have high humidity all year round and thus higher problems of damp and mould. For many years, condensation problems were blamed on the lifestyles of occupants, particularly in social housing, where excessive moisture came from such 'anti-social' activities as washing and drying clothes, bathing, boiling kettles and using the wrong kind of heaters! An important study carried out in Glasgow at a time when anti-social tenants were fair game in newspapers and blamed for problems of damp, concluded that it was the construction of concrete buildings and external factors that were more responsible for dampness, and high-rise, system-built blocks were at greater risk as they were exposed to driving rain off the sea. 'Vapour pressure maybe greater externally and moisture vapour can be forced into buildings'.[35]

Many years later in Glasgow, possible solutions to reducing dampness in houses from clothes drying were put forward by the Mackintosh Environmental Architecture Research Unit (MEARU), calling for the inclusion of dedicated low-energy drying facilities in houses, which are not currently a requirement of building regulations. People living in fuel poverty cannot afford to run tumble dryers and these, if used, may even be vented inside houses. The dampness caused by this may also lead to an increase in dust mites.[36]

Mould and damp problems are endemic in most buildings, with reports that over 10 million families are living in damp houses in the UK. The *Daily Mail* newspaper suggests that half the population live in damp houses.[37] However the extent of the problem has been questioned by RICS. Rather than being concerned about the health of building occupants, the interest of RICS in this subject was due to concern about the health and insurance effects on their members! RICS found hardly any examples of mould growth in the 1,200 properties they claim to have surveyed.

> Concerned about the possible legal and insurance effects on chartered surveyors who regularly investigate mould in peoples' homes, RICS commissioned research into the levels of mould in homes, hospitals, schools, offices and other dwellings across England.
>
> Only 1% of the research population were found to have Stachybotrys chartarum and/or Aspergillus fumigatus. Only 5.4% of the research population were found to have any significant mould growth at all.[38]

This is in complete contradiction to numerous academic studies of mould growth that have found that *all* homes studied contain mould cells. This is established by examining house dust scientifically.[39] [40] [41]

The team at Exeter University, which linked asthma with energy efficiency measures, also reviewed findings from 17 studies in eight different countries.

> Moulds are abundant in our outdoor and indoor environments, with around 10 varieties living in a typical home. We've found the strongest evidence yet of their potentially harmful effects, with higher levels of some of these moulds presenting a breathing hazard to people suffering from asthma, worsening their symptoms significantly. It also looks as though mould may help to trigger the development of asthma – although research in this area is still in its infancy. This research has highlighted the need for housing providers, residents and healthcare professionals to work together to assess the impact of housing interventions. We need to make sure that increasing the energy efficiency of people's homes doesn't increase their exposure to damp and mould, and potentially damage their health.[42]

There are a number of different species of mould including *Penicilium spp.*, *Aspergillus versicolor*, *Cladosporium sphaerospermum*, *Chaetomium globosum*, *Alternaria*, etc., and all have a significant effect on health.

> Epidemiological studies … support the significant link between mould exposure and certain pathologies, generally respiratory such as an increase of asthma, dyspnoea, wheezing, cough, or respiratory infections like rhinitis and bronchitis. Among affected populations, children are particularly at risk: wheeze and persistent cough in the first year of life and asthma development.[43]

There is substantial epidemiological evidence from studies conducted in many different countries and under different climatic conditions, to show that the occupants of damp or mouldy buildings, both houses and public buildings, are at increased risk of respiratory symptoms, respiratory infections and exacerbation of asthma.[44,45]

Unfortunately few studies could be found on the health impacts of improved conditions, but some studies are available showing that remediation of dampness problems can improve health outcomes. Medical studies rarely if ever provide useful technical information on the associated remedial measures that have been carried out.

> While groups such as atopic and allergic people are particularly susceptible to biological and chemical agents in damp indoor environments, adverse health effects have also been found in nonatopic populations. The increasing prevalences of asthma and allergies in many countries increase the number of people susceptible to the effects of dampness and mould in buildings [46]

A further problem in this area is the lack of scientific and medical research about the effects on mycotoxins. Mycotoxins are secondary metabolites that are produced by mould and can be much more dangerous to health but more research is needed to provide evidence of direct causal links between mould toxins and disease.[47,48,49] Mould toxins are also linked with VOCs.

> Several fungi produce volatile metabolites, which are a mixture of compounds that can be common to many species, although some also produce compounds that are genera- or species-specific. Microbial volatile organic compounds are often similar to common industrial chemicals. To date, more than 200 of these compounds derived from different fungi have been identified including various alcohols, aldehydes, ketones, terpenes, esters, aromatic compounds, amines and sulfur-containing compounds.[50]

Sources of moisture and relative humidity

> The most important means for avoiding adverse health effects is the prevention (or minimization) of persistent dampness and microbial growth on interior surfaces and in building structures. The prevalence of indoor damp is estimated to be in the order of 10–50%. It is highest in deprived neighbourhoods, where it often significantly exceeds the national average. Many case reports have also shown dampness and mould problems in office buildings, schools and day-care centres, but it is unclear what proportion of these buildings is affected.[51]

One of the main aims of building materials and fabric is to keep water out of buildings. Despite this simple principle, many buildings leak and suffer from water penetration and damp from floors, walls roofs and poorly maintained rainwater goods and plumbing.[52] Even when there are no leaks or defects, moisture is always present in buildings due to high levels of humidity, generated by human activities and the use of non-hygroscopic and plastic materials that cannot deal with condensation.

Moisture in the air comes from climatic conditions but also from the occupants of buildings and their activities. Figure 7.1 indicates how much water can typically be added to a house in a day by two people, including clothes drying.

'30 per cent of moisture in homes is attributable to clothes drying on wash days' [53] It is well understood that relative humidity in buildings should be between 40 per cent and 60 per cent. Air that is too humid or too dry can affect a variety of health conditions.

Materials, finishes and surfaces

Mould will grow on most materials if sufficient moisture is available, however the susceptibility of construction materials to mould can vary. Johannson *et al.* give the impression that natural materials are more susceptible to mould than synthetic materials but much of this

2 people at home can produce	=	3 pints
A bath or shower	=	2 pints
Drying clothes indoors	=	9 pints
Cooking and use of a kettle	=	6 pints
Washing dishes	=	2 pints
Bottled gas heater (8 hours use)	=	4 pints
Total moisture added in one day	=	26 pints or 14.8 litres

Figure 7.1 Water from human activities
Source: Liverpool Healthy Homes[54]

research is based on laboratory tests, which endeavour to simulate conditions in buildings.[55] More recent research into moisture issues in buildings tends to rely on computer modeling, sometimes based on data from laboratory climate studies. While much of this work appears to be useful, climate chambers do not easily simulate conditions in real buildings. Indeed the science of hygrothermal issues in buildings is still in its infancy and its importance has only been recognised in recent years. Building physics as a discipline has rather glossed over this important area with the exception of the work of Hens.[56] While Hens' work lays down many scientific principles for moisture transfer and both hygroscopic and non-hygroscopic materials, these principles are rarely applied in everyday practice and the formulae in Hens' book would be impenetrable to most readers.

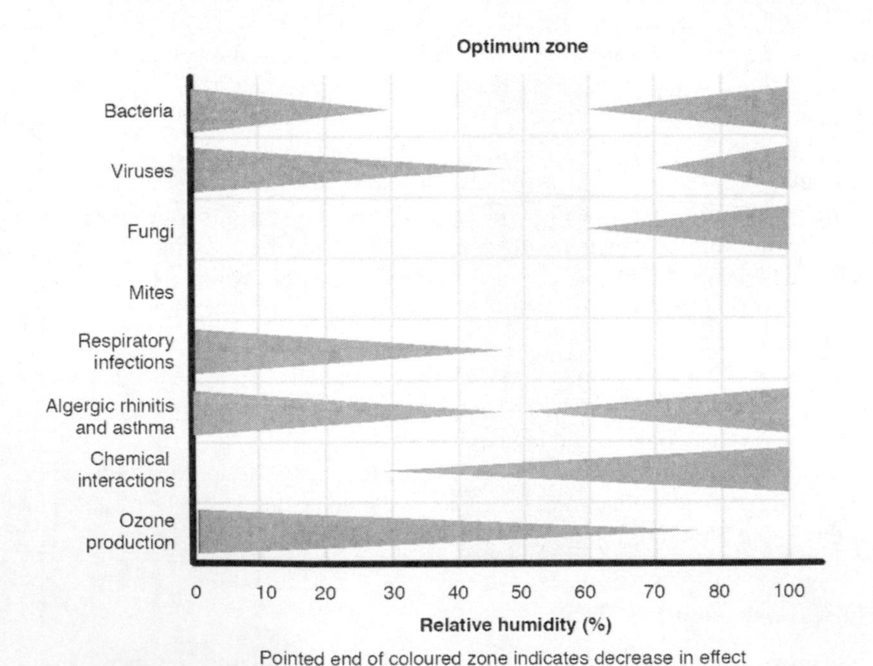

Figure 7.2 Critical relative humidity for health and illness

WHO has suggested that there should be 'limit values' for relative humidity on surfaces and structures applying principles of 'hygrothermal design' to limit mould growth. Humidity should be restricted below 75 per cent, though higher levels would be permitted for 'clean' materials.

> Even if the lowest relative humidity for germination of some species of fungi is 62–65%, experiments on common building and finishing materials indicate that susceptible surfaces can be kept free of fungal growth if the relative humidity is maintained below 75–80%. Mould fungi do not grow below a relative humidity of 80% or below 75% within a temperature range of 5–40°C.[57]

The Swedish government has been concerned enough about damp and mould to finance studies in this area, which have included some limited investigations of real buildings. Despite the perception that Sweden has high standard buildings, it does experience serious damp and mould problems. A study in Sweden found 751,000 dwellings with moisture problems. This was particularly severe in buildings built using the external thermal insulation composite (ETIC) construction system.[58]

In another study in a Swedish climate chamber, moisture levels in ten different building materials were examined for 12 weeks. These were also tested in three outdoor ventilated crawl spaces and attics of real buildings over two and a half years. This study found quite a wide variation in the susceptibility to dampness of similar materials and results seem inconclusive. Unsurprisingly manufacturers of building products do not refer to mould growth so specifiers and home owners have to exercise their own judgement about what materials might be selected to minimise mould risks.

The list of materials used in the Swedish susceptibility to dampness test[59] include:

- cement-based board 8 mm
- EPS insulation board 50 mm
- rigid glass wool insulation board 15 mm
- asphalt paper 1.5 mm
- wet-room gypsum plasterboard
- gypsum board with cardboard surfaces 13 mm
- plywood 12 mm (softwood plywood)
- thin hardboard 3.2 mm
- particle board 12 mm
- pine sapwood tongued and grooved board 19 mm

Table 7.1 Critical relative humidity range for various groups of materials

Material	*Relative humidity (%)*
Wood and wood-based materials	75–80
Paper on plasterboard	80–85
Mineral insulation materials	90–95
Extruded and expanded polystyrene	90–95
Concrete	90–95

Source: Johannson[60]

Gypsum plasters and lime

There are some anecdotal references to gypsum plasters and plasterboards being particularly susceptible to mould growth. Thus some research has been carried out to test the use of biocides with gypsum, to inhibit mould growth, but these have had limited success.[61]

During the 1960s gypsum plasters, particularly lightweight plasters mixed with vermiculite, were adopted to replace lime and more traditional plasters, and were an attempt to achieve quicker setting times. However gypsum plasters could still be slow to dry out and often walls were painted with acrylic paints before the plaster had fully dried, trapping moisture. Gypsum plasters were soon found to cause problems with older properties as salts and mould would appear. Some gypsum plasters may limit vapour permeability and this might trap dampness. Gypsum plasters can also fail when wet, due to leaks or flooding. Metal angle beads, conduit buried in walls, etc. can rust due to the moisture trapped in gypsum plasterwork.[62]

Due to these problems so-called renovating plasters were introduced, many of them lime based, rather than using gypsum. Lime plasters involve reverting to traditional methods, long used in the renovation of historical buildings. For many years lime was considered marginal to mainstream construction and merely a specialised solution for historic buildings. But in more recent times a range of lime and lime hybrid products has been increasingly used in new buildings as well as in renovation. Mainstream companies tend to market blends of cement and lime on a 50/50 basis, as this is cheaper and more attractive to mainstream building contractors. Cement lime sand plaster (often used with perlite) is said to be lighter and more breathable than gypsum or a straight cement sand mix but would be regarded by lime experts as an unsatisfactory compromise.[63]

Whilst the importance of using lime has been recognised, some builders use inferior materials by adding hydrated lime to a standard cement and sand mix and then claiming it is a lime plaster. Cement external renders (even with some lime added) still have a tendency to trap moisture in walls and this can rapidly lead to mould growth. Other approaches retain gypsum and cement based plasters but add water inhibitors and other chemical compounds including fungicides and anti-mould paints.[64]

There is an extensive range of plastering formulations and care must be taken to ensure that the right materials are used to maintain vapour permeability and reduce dampness problems. Using lime plasters rather than cement based materials is preferred by some. Lime mortars, plaster and renders are useful materials in efforts to create healthier buildings because lime is breathable (vapour permeable). Thus walls built with lime will have less tendency to trap dampness. Additionally, lime is a biocide and this makes it much harder for mould growth to become established. 'The alkalinity deters wood-boring beetles and helps sterilise walls.'[65]

For centuries, lime-washing walls was a standard 'paint' on stone, brick and mud walls. Stables, dairies and food processing buildings were treated with lime-wash, though this is unlikely to be accepted by environmental health inspectors today! Lime-based products are sometimes used today as a primer to deal with mould problems instead of chemical-based biocides. Lime and lime-based renders were sometimes treated with linseed oil or tallow but this may reduce its biocidal effects.

> In most parts of Wales, stone buildings were traditionally limewashed externally as a matter of course for practical, as well as aesthetic, reasons. It provides a protective coating, which acts as a barrier against penetrating damp. While providing a good

weatherproof cover, limewash is also permeable, allowing any water within the structure to evaporate through the surface. It does, however, need to be applied regularly. Binders, such as casein and tallow can be added to improve adhesion and water shedding, but these can sometimes encourage mould growth unless a biocide is added. They also reduce permeability, which may make them unsuitable in some situations.[66]

Casein paints have been popular as an ecological wall finish. Casein is a protein found in milk, and some natural building books offer recipes for natural paints based on milk. This can lead to mould problems however, but there are commercially produced lime and casein mixed paints and finishes available. Certain manufacturers claim that casein paints will help to regulate humidity. If chemical fungicides have been added to casein paints this rather negates their health benefits.

Modern lime plasters and renders are often made with natural hydraulic lime (NHL) but hydrated lime and lime putty are also used. Lime products are caustic and should be handled with care when used in buildings, but they are not toxic and do not have any negative health impacts once in a building. Lime is calcium oxide or hydroxide and gradually carbonates over time, reverting to its original form.[67]

The correct use and variety of lime building materials is a broad subject but there is plenty of advice available from the UK Building Limes Forum[68] and a range of books on lime and natural materials.[69,70,71]

Hygroscopic materials: insulation and finishes

It was assumed for many years that hygroscopic materials in buildings were bad as they would attract damp and then rot. This is the basis of much of the prejudice against natural, renewable and ecological materials. Materials with hygroscopic characteristics do attract moisture, and if used in the wrong way and the wrong place, they will become damp and may start to break down. However many petrochemical synthetic materials also fail when they get wet so this is not just a problem for natural materials.

Strawboard products in the 1970s and 1980s were used in many low-cost housing developments, but were sealed into damp cavities and began to rot. Strawboard partitions were even used in poorly designed flat roofs and bathrooms where they soon got damp and failed and they were also used in ceiling panels.[72] But in recent years strawboard has made a come back as natural materials are better understood and used correctly in breathable constructions.[73]

It has now been recognized that hygroscopic materials have significant benefits in buildings by assisting in managing moisture and humidity. If a material can absorb moisture in either the fabric or interior of a building, without being damaged, this will make it possible to maintain relative humidity at the ideal 40–60 per cent. Non-hygroscopic materials are unable to do this and are thus much more prone to condensation forming on cold surfaces.

Hygroscopic materials include boards and insulations such as sheep's wool, wood fibre, hemp and flax fibres and hemp lime. This can be supplemented through the use of breathable lime and clay plasters, renders and finishes. Walls made from earth, clay and strawbales can also perform a similar function. The potential value of this was recognised recently by the UK government's Department of Energy and Climate Change (DECC). DECC issued a consultation document on moisture risk assessment. This consultation was prompted by an organisation called the Sustainable Traditional Building Alliance (STBA) that has been campaigning for a new approach to the renovation and retrofitting of old buildings using methods and materials that are more sympathetic to existing fabric.

> The purpose of this new guidance is to provide the backbone of an integrated and holistic approach to moisture risk throughout the design, construction, alteration, repair, maintenance and use of buildings. The desired outcome is reduced risk, through an informed process and from a solid understanding of the underlying building physics. The scope of the guidance is work on all buildings, both new and existing, and all building elements.[74]

> Moisture buffering can give a degree of protection against short-term moisture problems particularly where there is a high degree of airtightness and hygroscopic materials with rapid response are available at or near the inner surface of the room. Furthermore, biogenic materials can have positive effects on the durability and health of buildings in regard to their active hygroscopic, capillary and vapour permeability qualities.[75]

> Modern framed walls, with either timber or steel framing, usually incorporate a 'sheathing' layer on the frame to provide racking resistance to the structure. It is typically a moisture-closed material and if it is outside the insulation layer may provide a barrier to vapour diffusion; this will lead to a risk of interstitial condensation …
>
> Moisture-open constructions often have the sheathing layer on the inside of the insulation and use moisture open insulations and membranes.[76]

The DECC documents carefully avoid referring to specific materials or material types, and create a certain amount of confusion about hygroscopic materials, moisture buffering and moisture open construction. The term 'moisture open' is confusing as normally 'vapour open' is the conventional term. Having commissioned work and issued consultation documents in 2014, this issue appears to have been quietly buried by DECC and nothing could be found about the outcome of the consultation. Having questioned government officials responsible for this work, it became clear that lobbying by industry groups had effectively sidelined the issue, which was now going to be the responsibility of 'industry' rather than a matter for government policy.

STBA, by contrast, has published a number of case studies of examples of buildings that have been retrofitted, comparing the use of synthetic insulation (in this case PIR) with wood fibre and lime plaster. Their research shows that the PIR insulated wall created an 'upward trend' with relative humidity (RH), whereas the wood fibre did not raise the risk of interstitial moisture and performed a buffering role.

> House 1 – Drewstaignton 600mm thick granite walls. 100mm PIR insulation was used. Walls demonstrate an upward trend in Relative Humidity and are above 80% – favourable conditions for mould growth,

> House 2 – Brick end of terrace, Shrewsbury with a different orientation to the Drewstaignton house – Handmade soft brick, 345mm thick which is much more porous as it has a more open structure. Internal insulation was woodfibre covered with a lime plaster. Dewpoint margins continue to be separate with fairly steady relative humidity (RH) over the year. Continues to perform well with low risk of interstitial moisture. Monitoring shows that the insulation appears to be performing a buffering role with respect to temperature and humidity.[77]

Research into the use of hygroscopic, natural breathable materials is still at an early stage and there may not be enough hard evidence to convince skeptics. Historic Scotland has carried out some important comparative research into retrofitting insulation in old masonry buildings. Hemp boards, wood fiberboards, blown cellulose, aerogel board and bonded polystyrene bead were used in a 19th-century tenement building.

> The insulation trials described in this case study show that six different internal insulation measures could significantly improve the thermal performance of mass masonry walls. Four of these measures achieved a U-value of below 0.3 (W/m^2). All of the measures were aimed at maintaining vapour permeability, and monitoring of moisture levels at the interface between the insulation and the masonry and at 50 mm within the masonry show that there are no significant increases in moisture due to the insulation work. Further on-site testing of internal insulation measures are on-going and the results of these, along with those from other Historic Scotland pilot studies show that the thermal performance of traditionally constructed mass masonry walls can be significantly improved using materials which are sympathetic to the existing building fabric.[78]

Historic Scotland *Technical Paper 17* considers these in more detail.

Table 7.2 Historic Scotland tests on insulation materials

Insulation type	*Average relative humidity of room (%)*	*Average relative humidity at interface between wall and insulation (%)*	*Average relative humidity 50mm into the wall fabric (%)*
100mm Hemp board	52.1	65.2	66.6
80mm Wood fibreboard	20.7	61.7	58.9
40mm Aerogel board	No results	No results	No results
50mm Blown cellulose	21.9	14.8	14.3
50mm Aerogel board	45.9	64.4	63.3
50mm Bonded polystyrene bead	58.3	16.4	15.8

Source: Historic Scotland[79]

Moisture buffering using hygroscopic materials

The ability of hygroscopic materials to control humidity in a room or a building is sometimes referred to as 'buffering'. While there are materials that can be very effective in this respect, very often the hygroscopic properties are inhibited through the use of non-breathable finishes. A simple material such as plasterboard (drywall/gypsum board) has some hygroscopic properties but when plastered and painted with conventional acrylic paints, these become negligible.[80]

It is possible to use finishes and paints both internally and externally that are 'breathable' but even when specified these are frequently substituted by builders who do not understand the difference. Synthetic plastic-based paints, for instance, will have a high vapour resistance. Research at the Fraunhofer Institute has shown that linings with a moisture buffering capacity had a significantly better ability to have a beneficial effect on IAQ when compared with conventional plaster finishes. They tested different kinds of timber panels, boards and logs and insulation from wood fibre and cellulose. The 'sorptive' ability of the insulation materials was diminished by the plasterboard and other linings however. Wood fibreboard gave the best results with an 80 per cent reduction in moisture, better than solid wood and other materials.[81]

A Nordtest report on moisture buffering from Denmark involved a wide range of partners from Norway, Sweden and Finland in a workshop and testing in 2003 using climate chambers/climatic rooms.

> The Workshop concluded that the phenomenon (of moisture buffering) 'is real' and an issue of public and industrial interest and debate – although the appraisal of moisture buffering materials has so far been based on a variety of definitions, so manufacturers could choose the one definition that seemed most suitable for a given cause.[82]

Disappointingly the tests did not evaluate hygroscopic materials, instead evaluating concrete, gypsum, brick and wooden panels, and it is hard to interpret the results as the aim was to establish testing protocols and methods.

Tests of moisture buffering at the Oak Ridge Laboratory in the USA evaluated wood panelling, gypsum and wood fibreboard, also gypsum board with cellulose and mineral wool insulation. An interesting result was that the use of a 'permeable paint' still significantly reduced the buffering capacity of the materials.

> The results show that building materials exposed to indoor air can have a strong effect on the indoor air humidity. Potentials, practical applications, and design concepts for utilizing the moisture-buffering effect of building materials are discussed. Results show that it is possible to design a permeable envelope with good moisture performance. In fact, a permeable envelope made of hygroscopic materials is less susceptible to condensation and mold growth at the internal surface of thermal bridges because the peak indoor humidity is lower when applying hygroscopic materials.[83]

Some natural materials have been put forward as having exceptional moisture buffering capacity. Unfired earth, in particular, can be over 20 times better than cellular concrete in terms of moisture buffering, according to Padfield.

> The revelation of the good moisture absorption of unfired clay brick will not come as a surprise to the large fraction of mankind which lives in earthen houses, but this article will remind engineers and architects in rich nations of a hidden merit of a cheap and unfashionable building material.[84]

An experimental low-energy house was built in Scotland in 2003, by architect Tom Morton, to study the use of unfired earth. Instead of using airtightness membranes, the walls incorporated unfired earth bricks together with clay plasters. A considerable amount of scientific monitoring and evaluation of the house was carried out. 'It was demonstrated that no condensation occurred in the fabric of the building. The regulation of air relative humidity was also confirmed by the monitoring, with internal air RH generally contained between the intended limits of 40 and 60%'.[85,86]

From the limited number of studies carried out on moisture buffering it is clear that hygroscopic materials are most successful when in direct contact with indoor air. If materials such as sheep wool insulation is concealed behind plasterboard and vapour-retarding layers then there is scarcely any effect on humidity levels.

> The only way to include the insulation as a significant influence on the interior climate is to use a porous paint on the plaster board and to move the vapour retarder towards the middle of the insulation. This idea is controversial, because there is a risk of condensation within the wall in cold weather. The constant temperature experiments reported here can shed no light on this matter.[87]

Hygrothermal issues and 'robust' construction

In order to understand damp and mould issues and ways to reduce these risks, much more should be known about hygrothermal issues in buildings. While UK building regulations refer to so-called 'robust details' (that have more recently become known as accredited details), they do not explicitly deal with hygrothermal performance. Robust or accredited details are concerned in part with avoiding 'cold bridging' (thermal bridging) and ensuring continuity of insulation. Cold bridging is important as it leads to cold spots, which is where condensation and mould growth is most likely to occur.[88]

There are some relevant ISO/British standards dealing with hygrothermal issues, but they are limited in what they cover. For instance BS/EN/ISO 13788:2002 does refer briefly to hygroscopic materials but only assesses condensation risk on surfaces.

> The assessment of the risk of interstitial condensation due to water vapour diffusion. The method used assumes built-in water has dried out and does not take account of a number of important physical phenomena including:
>
> - the dependence of thermal conductivity on moisture content;
> - the release and absorption of latent heat;
> - the variation of material properties with moisture content;
> - capillary suction and liquid moisture transfer within materials;
> - air movement through cracks or within air spaces;
> - the hygroscopic moisture capacity of materials.
>
> Consequently the method is applicable only to structures where these effects are negligible.[89,90]

Crucial issues such as the vapour permeability of materials and so-called breather membranes are long overdue for review as standards were written in the 1990s, long before super insulating buildings became the norm.[91,92] As buildings become more heavily insulated, predicting condensation risks is very important, but this is primarily done through computer simulation models.[93]

There are several hygrothermal simulation model tools available and university experts use such tools to predict moisture risk. But there is no certainty that such computer models are an adequate substitute for good building practice and detailing. Ideally prediction calculations should dovetail with good detailing but in the current state of practice this is far from the case as there is insufficient communication between practicing designers and the academic scientists. There is insufficient feedback based on post-occupancy evaluation of where things have gone wrong and condensation and mould growth have appeared. For instance airtight forms of construction, where elements and insulation are sealed into walls and roofs, without any ventilation or vapour permeability to allow moisture to escape, are fraught with danger but knowledge of these problems is mainly anecdotal. Robust or accredited details and the building regulations will not give enough guidance to avoid dampness and condensation and do not provide adequate tools to predict hygrothermal performance. Computer prediction tools known as condensation risk analyses (CRAs) may not give realistic assessments of what will actually happen in real buildings, but this is also explained in terms of bad building practice as suggested in a study in 2011 by Leeds Metropolitan University and University College London.

> The findings of the fieldwork suggest that the house construction industry is not able to produce, with any degree of consistency, construction that is well designed and achieves the performance required of robust construction.
>
> It is evident that the influence of workmanship on moisture performance exists, and can have a significant effect on degrading the hygrothermal performance of 'as built' robust details.
>
> Although no surface condensation risk is predicted for the given conditions, 10 out of 12 'as built' details failed to meet the standard requirements regarding mould growth.[94]

Conventional practice and guidance from the building regulations is often contradictory and misleading. Sometimes there is confusion in the advice about the difference between impermeable membranes and breathable membranes. Both may be used with little understanding of the degrees of vapour permeability that will be achieved, as different kinds of membranes have different degrees of vapour permeability. There is a rule of thumb that materials in wall construction should become more vapour open as they get closer to the outside, on the basis that this will ensure vapour movement from inside to outside. Most commercial product manufacturers make claims about breathability, but even if correct, these only apply to one layer in what is usually a multi-layer build up of materials. 'Breather' membranes are used in roof and wall construction, as standard practice today, but prediction of how moisture will actually behave is not yet an exact science.

Awareness of these problems is growing. There will be a conference in Florida in the USA, organised by the American Society for Testing and Materials (ASTM) in October 2016, to discuss Advances in the Hygrothermal Performance of Building Envelopes: Materials Systems and Simulations.[95]

It is vitally important that computer scientists developing simulation models, base these on actual experience of buildings and how they perform.

How water vapour behaves: a new approach to building physics

The warmer the air, the greater the amount of water (as vapour) air can contain. Water is constantly moving from its liquid state to vapour and vice versa, not just when condensation can be observed. When discussing this it is necessary to understand the difference between absorption and adsorption. *Absorption* is the movement of liquid water into the pores of both hygroscopic and hydrophilic materials. Hydrophilic materials will absorb much less water. *Adsorption* is when water forms a thin film on the surface of a solid (internally or externally). The reverse process when a material releases water is called desorption.

The movement of water within a material when absorbed involves energy and thus the diffusion of moisture through a material is influenced by heat and RH. As RH rises then moisture content (MC) will change. Hygroscopic materials can absorb more water than non-hygroscopic materials but the MC will always be changing and vapour will be exchanged with the surrounding air. Understanding the building physics of moisture in buildings in this way goes against conventional wisdom, which has generally been to use hydrophilic materials in an attempt to repel rather than absorb moisture. The ability of the indoor environment and the hygroscopicity of materials will determine how effectively mould formation can be inhibited, whereas hydrophilic materials, that cannot absorb moisture, may lead to much greater problems of mould growth. Providing hygroscopic materials are also durable and robust, they can be a much more effective way to control damp and mould.

An even more surprising issue in building physics is the ability of moisture contained within hygroscopic materials to improve their thermal performance, as water can retain heat through its thermal mass. Thermal mass effects cannot be fully understood without also considering moisture issues. Concrete, which is believed by many architects and engineers to help to heat and cool buildings through its thermal mass, is not as effective as hygroscopic materials. The thermal lag effect, known as decrement delay (the time it takes for heat to pass through a material), is rarely considered when designing buildings. It is suggested that the thermal mass of concrete is not as effective as claimed as it is not sufficiently dynamic to respond to changing conditions. Also while concrete can absorb some moisture, cement and concrete surfaces tend to adsorb rather than absorb, with moisture staying on the surface. This can lead to mould growth.

Natural hygroscopic materials such as wood fibre, hemp lime and unfired earth are much more effective thermally in terms of their thermal mass because of what some refer to as 'phase change' capacity. Phase change simply refers to the process of water changing its state from say vapour to liquid. Insulation materials that are claimed to both store and release energy, as well as resisting its transfer, create novel possibilities. Most conventional synthetic lightweight insulations containing trapped air have almost no phase change possibilities, but natural hygroscopic materials do, as the water they absorb can change from liquid to vapour within the material. However using the term phase change can be confusing and it is poorly understood within the construction industry.

Vapour permeability and breathability

The word breathable has become part of construction language in recent years but is widely misunderstood and best avoided for scientific accuracy. However the contrast between buildings that can 'breathe' and those that are tightly sealed up with synthetic materials is relatively easy to understand and gives some idea of the difference in approaches between healthy and less-healthy buildings, hence the subtitle to this book *No Breathing Space?*

Good IAQ requires fresh air from good ventilation, as discussed in Chapter 8, but as set out in this chapter, good control of moisture and humidity is also dependent on using hygroscopic and vapour-permeable materials. It is important to discover the vapour permeability of construction materials before they are specified to ensure that they will ensure good air quality.

Many standard construction materials have very poor vapour permeability or make exaggerated claims about their vapour performance. It is not uncommon to find some building products advertised as both moisture resistant and vapour permeable. This is not necessarily a contradiction, but there is a significant variation in how certain materials cope with absorbing or adsorbing moisture.

There are three main forms of building product where vapour permeability is critical: (1) vapour control/vapour check/vapour barrier membranes; (2) construction boards such as gypsum/plaster boards and sheathing boards, such as plywood and OSB; and (3) standard construction materials such as plastered and rendered brick and block, etc.

The m-value or mu-value of a material is also known as its 'water vapour resistance factor'. It is a measure of the material's relative reluctance to let water vapour pass through, and is measured in comparison to the properties of air. The m-value is a property of the bulk material and needs to be multiplied by the material's thickness when used in a particular construction.

Because the m-value is a relative quantity, it is just expressed as a number (it has no units). However vapour resistance is expressed at MNs/g. For example to calculate the vapour resistance of a material with a mu-value of 4000 and thickness of 3 mm, the vapour resistance is,

$$4000 \times 0.003 \text{ m divided by } 0.2 \text{ g.m/MN.s} = 60 \text{ MNs/g}$$

Energy efficiency software tools should include vapour resistance calculations but it is not at all clear how effective these are related to the energy efficiency calculations as the main concern is achieving airtightness.[96,97]

It is uncommon to find vapour permeability figures quoted by manufacturers as the industry is mainly concerned with vapour resistance. Even if such figures are available, architects and specifiers may not know how to use them. However the Table 7.3 of vapour resistance values is useful for comparison purposes.

Information about vapour resistance and breathability is presented on product data sheets in so many different ways, using different forms of measurement, that it makes life very difficult for the designer or specifier. The figures quoted in Table 7.3 should be seen as a guide only. Without understanding the basis of MNs/gm, the comparison between different figures is useful. Similar products and r values from different manufacturers may be quite different. 'Little information is available about the comparable properties of commercially available materials or what to consider when selecting the appropriate product for a particular application. Both building code requirements and *vendors' product information are inconsistent and confusing*'.[98]

UK building regulations provide guidance on the use of 'breathable' building felts and membranes in roofs but the primary concern is with barriers to the passage of air to achieve airtightness standards. Humidity is regarded as being controlled by ventilation rather than the use of hygroscopic or vapour permeable materials. Building regulations in the future will need to be changed to take account of hygroscopic and vapour permeable materials.

'Breather membranes' are normally used in roof construction and generally in timber-frame wall construction. The breathability of these membranes varies enormously and this may not be taken into account in condensation risk analyses. If the vapour permeability is not specified then a much more vapour-resistant membrane may be used. There is need for much more research into the performance of these barriers. Due to the preoccupation with airtightness barriers, industry has tried to come up with membranes that restrict air movement (air resistance) and water resistance to prevent any water ingress through external cladding, but materials are also required to be vapour permeable in an attempt to prevent interstitial condensation.

In the USA vapour permeability is measured using ASTM E 96, which calculates how much vapour can pass through a membrane or material in 24 hours and this is expressed in

Table 7.3 Typical 'r' (vapour resistance) values

Material	*Typical resistivity (r)*	*MNs/gm*
Air		5
Cement plaster	100	
Lime plaster		75
Clay plaster		40
Gypsum plaster		50
Synthetic plasters		1,500
Concrete varies by density		50–500
Aerated concrete	35	
Clinker blocks	150	
Bricks		50
Gypsum (plaster) boards/dry wall		60
Plasterboard foil backed		4,800
Clay boards		90
Strawboard		60
Wood-wool		35
Expanded polystyrene		150
Extruded polystyrene	1,000	
Polyurethane foam		300
Poly foam with foil		10,000
Polyiso foam with foil		43,000
Polyethylene sheet		120
Fibre insulations		6
Wood fibreboards		25
Hardboard		500
Cellulose insulation		45
Softwood		200
Hardwood		400
Plywood		500
OSB		200
Indoor plastic emulsion paints		1,500
External plastic paints	15,000	
Silicate paints	300	
Lime-wash (5 coats)		250
Solvent-based gloss paints		20,000

Source: Based on information from Neil May, *The Engineering Toolbox*, CIBSE (1999), Butt (2005) and various product data sheets including Keim, Pavatex, etc.[99,100,101]

'perms'. A range of membranes or 'housewraps' have been developed, from a building paper to so-called 'smart membranes' that are made from various plastic materials such as polyethylene, polypropylene and polyolefins. Various products can range from 60 perms to as low as 5 perms. However, in the UK it is more common to provide vapour resistance figures for products such as 0.25–0.6 MNs/g. There is confusion between air permeability and vapour permeability in the information provided by manufacturers. The National House Building Council (NHBC) lacks confidence in what it calls VPUs (vapour permeable roof underlays) and recommends the need for ventilation in cold roofs even where VPUs have been certified to avoid condensation.[102]

Local authority building control guidance sometimes refers to a brand name product 'Tyvek' (or similar and approved) but there are many other products available. Some manufacturers claim that their membranes are 'smart' or 'intelligent', in that they claim they can adapt to different climatic conditions or perform in a dynamic way. There is even an Intelligent Membrane Association.[103] It was not possible to find any independent scientific evaluation of the smart and intelligent claims other than the normal trade certification through British Board of Agrement. One manufacturer, for instance, says that the diffusion resistance of their product will change as humidity varies and that it will be more vapour open in the summer when the weather is hotter. Membranes also rely heavily on sticky tape to ensure that all joints are sealed to achieve the required level of airtightness, but there is some doubt as to whether the synthetic adhesives used will last as long as the membrane materials.

While it seems hard to avoid the use of synthetic membranes in roofs, their use in walls is much more questionable. In timber-frame construction, it is common to use timber composite boards such as OSB as an airtightness layer and then a 'breather' membrane towards the outside of the wall. It is not uncommon to see non-breathable plastic layers used in timber-frame construction.

Some construction books refer to moisture gradients (sometimes referred to by the letters DY) in walls. Some show a correlation with the temperature gradient. The basic principle that has been applied for many years is to make the materials on the inside of a building as vapour resistant as possible and to increase permeability to the outside. This is based on the commonly held idea that vapour will always move through the layers towards the outside, passing from the less vapour permeable layers to those that are more permeable. While this can take place if there is a vapour pressure gradient, vapour will not automatically move towards colder air. This remains an area of debate, with some advising the use of vapour impermeable surfaces internally, whilst others advise quite the opposite. Interstitial condensation can occur if there is warm moist air that can condense onto a colder surface and this often occurs *within* the various layers of both masonry and timber-frame construction (referred to as the dew point). Such condensation can lead to rot or decay within the fabric.

The standard method for predicting interstitial condensation in lightweight constructions is known as the Glaser method (BS5250/EN 13788) but this has been criticised as failing to take account of dynamic moisture movements. 'The Glaser method averages everything to a monthly calculation and is "steady state". In particular it omits driving rain from its calculations which is probably the most significant of all factors in the failure of all insulation systems on all buildings'.[104] An alternative method is set out in EN/ISO15026 (2007), which is a dynamic model.

> Transient hygrothermal simulation provides more detailed and accurate information on the risk of moisture problems within building components and on the design of remedial treatment. While the Glaser method considers only steady-state conduction of heat and vapour diffusion, the transient models covered in this standard take account of heat and moisture storage, latent heat effects, and liquid and convective transport under realistic boundary and initial conditions. The application of such models has become widely used in building practice in recent years, resulting in a significant improvement in the accuracy and reproducibility of hygrothermal simulation.[105]

Vapour permeability under attack

It is argued by some that if hygroscopic moisture buffering materials are used properly in a building, and there is sufficient vapour permeability that allows moisture to escape or diffuse, this should reduce the reliance on ventilation systems to remove moist and bad air. This is a controversial issue and manufacturers of synthetic non-vapour permeable materials argue that 'breathability is a red herring', as good IAQ can only be maintained through ventilation (usually mechanical ventilation). Kingspan, one of the largest UK manufacturers of synthetic foam insulations, argues that 'breathable constructions and the breathability of insulation products are at best a side show and in reality they are a complete red herring in the avoidance of surface condensation, mould growth'.[106] Kingspan argues that bulk air movement through ventilation is the way to ensure good IAQ and it criticises advocates of breathability and natural materials. It also attacks what it claims are false claims from the mineral fibre insulation industry about the importance of breathability.

The Red Herring Report was based on work commissioned by Kingspan from the Cambridge Architectural Research Group (CAR). Their criticisms of breathing walls are based on a case study of cellulose-insulated timber-frame buildings at Findhorn and research done by Crowther and Baker as long ago as 1993. Breathing construction has advanced by leaps and bounds since then so CAR's conclusions about breathable construction should be treated with some scepticism. They state that:

> Air movement through 'breathable' panels will not attain levels sufficient to transport away moisture that has condensed within them due to poor design, or to remedy water ingress due to building failure. The key to creating and maintaining comfortable and healthy indoor conditions lies in good thermal design, linked with adequate provision for ventilation through controllable routes.[107]

CAR's conclusions illustrate a common misconception of breathable construction: that breathability is the same as air movement. The use of natural and hygroscopic materials, they say, cannot deal with a build up of condensation as they do not transport away moisture. But this is not what breathable materials do. Instead they reduce the risk of moisture build up through absorption and their hygroscopic properties. Vapour permeable materials do not transport moisture through air movement as they enable vapour to disperse. As May pointed out in a critique of the Red Herring Report 'It is irresponsible not only in regard to fabric health and human health, but also in regard to the positive possibilities of producing safer, healthier and better performing buildings'.[108]

VTT in Finland, another reputable scientific research body, has also provided confusing information on breathability. Despite giving a reasonably balanced explanation of the benefits of breathable and hygroscopic materials, in a research report they nevertheless came to the conclusion that 'a non-capacitative thermal insulation (polyurethane) could have significantly higher (about 100%) moisture buffering capacity than thermal insulation'. The report was 'requested by PU Europe', the trade body for the manufacture of polyurethane insulation! This study has been referred to by Kingspan and others as providing evidence that non-permeable insulation materials are 'breathable'.[109] Such apparently deliberate attempts to seed confusion about the science of breathability and permeability may lead some to ignore the potential benefits of using hygroscopic materials. Understanding of vapour transfer in buildings needs to improve significantly and be based on good science, not propaganda from synthetic materials companies.

Further discussion of the use of hygroscopic materials is included in Chapter 11, indicating a range of materials than can be used to reduce the risks of condensation and mould growth.

Notes

1 Poll conducted for the energy Saving Trust (2014), IPSOS www.ipsos-mori.com/Assets/Docs/Polls/ipsos-mori-energy-saving-trust-topline-august-2014.pdf [viewed 26.2.16]
2 Department of Energy and Climate Change (2015) *Annual Fuel Poverty Statistics Report*, DECC
3 Howieson, S. (2005) *Housing and Asthma*, Spon Press, London and New York
4 McCabe, J., *Inside Housing* 16.10.15, 28–30
5 Cavity Insulation Victims Alliance www.civalli.com/what-we-do/ [viewed 26.2.16]
6 Damp in social housing blamed on 'inappropriate' insulation, BBC News www.bbc.co.uk/news/uk-wales-politics-34081718 [viewed 26.2.16]
7 Cavity-wall insulation crisis may hit three million homes, *The Telegraph* www.telegraph.co.uk/finance/property/11485758/Cavity-wall-insulation-crisis-may-hit-three-million-homes.html [viewed 26.2.16]
8 Cavity Claim UK www.cavityclaimuk.com/ [viewed 26.2.16]
9 https://ciga.co.uk/ [viewed 26.2.16]
10 Inbuilt Ltd, Davis Langdon (2010) Study on hard to fill cavity walls in domestic dwellings in Great Britain, DECC ref: CESA EE0211, Inbuilt ref: 2579-1-1
11 Cavity wall, Energy Saving Trust www.energysavingtrust.org.uk/domestic/cavity-wall [viewed 26.2.16]
12 www.energysavingtrust.org.uk/our-experts [viewed 26.2.16]
13 Hansford, P. (2015) *Solid Wall Insulation, Unlocking Demand and Driving Up Standards A report to the Green Construction Board and Government by the Chief Construction Adviser*, Department for Business, Innovation and Skills
14 Hansford, P. (2015) op cit.
15 Loft and cavity wall insulation causes damp, Heritage House www.heritage-house.org/insulation-causes-damp.html [viewed 29.2.16]
16 Sharpe, R.A., Thornton, C.R., Vasilis Nikolaou, V. and Osborne, N.J. (2015) Higher energy efficient homes are associated with increased risk of doctor diagnosed asthma in a UK subpopulation, *Environment International* 75, 234–244
17 Boardman, B. (2009) *Fixing Fuel Poverty: Challenges and Solutions*, Earthscan, London
18 The Standard Assessment Procedure www.gov.uk/guidance/standard-assessment-procedure [viewed 29.2.16]
19 Kelly, S. (2013) Decarbonising the English Residential Sector, Modelling Policies, Technologies and Behaviour within a Heterogeneous Building Stock, PhD, Dept. of Land Economy University of Cambridge
20 Hills, J. (2012) *Getting the Measure of Fuel Poverty: Final Report of the Fuel Poverty Review*, Case Report 72 March 2012 DECC, London
21 Why is there condensation in my loft?, *The Telegraph* www.telegraph.co.uk/finance/property/advice/11333991/Why-is-there-condensation-in-my-loft.html [viewed 26.2.16]
22 Clark V. Affinity Sutton Homes Ltd. Barnet County Court 4 April 2014. http://nearlylegal.co.uk/2014/04/disrepair-damp-and-quantum/ [viewed 26.2.16]
23 Loft and cavity wall insulation causes damp, Heritage House www.heritage-house.org/insulation-causes-damp.html [viewed 26.2.16]
24 Reports of damp soar in social housing as residents avoid turning on heating, *The Guardian* www.theguardian.com/society/2013/dec/27/damp-social-housing-residents-heating-energy-bills [viewed 26.2.16]
25 Damp has ruined everything, says housing association tenant, News Shopper www.newsshopper.co.uk/news/9518308.Damp_has_ruined_everything__says_housing_association_tenant/ [viewed 26.2.16]

26 Sutton mum-of-three battles with housing association over mould, *Sutton Guardian* www.suttonguardian.co.uk/news/8456401.Mum_of_three_battles_with_housing_association_over_mould/ [viewed 26.2.16]

27 Borehamwood mother's mould-ridden house is 'depressing' to live in, *Borehamwood Times* www.borehamwoodtimes.co.uk/news/11682225.Mould_ridden_house_is_depressing_to_live_in_/?ref=mr [viewed 26.2.16]

28 Futurefit, *Final Report: An insight into the retrofit challenge for social housing* www.affinity-sutton.com/media/667531/future-fit-final-report-part-two.pdf [viewed 26.2.16]

29 Zero Carbon Hub (2014) *Closing the gap between design and as-built performance arch*, Zero Carbon Hub

30 Technology Strategy Board (2014) *Retrofit for the Future A Guide to Making Retrofit Work*, T14/026

31 Gorse, C., Glew, D., Miles-Shenton, D. and Farmer, D. (2015) Addressing the thermal performance gap: Possible performance control tools for the construction manager In Raidén, A.B. and Aboagye-Nimo, E. (eds) *Procs 31st Annual ARCOM Conference*, 7–9 September 2015, Lincoln, UK, Association of Researchers in Construction Management, 337–346

32 Riversong HouseWright https://riversonghousewright.wordpress.com/about/9-hygro-thermal/ [viewed 26.2.16]

33 Brocklebank, I. (2008) Magic wallpaper and the problem with dew http://ihbc.org.uk/context_archive/103/broklebank/page.html [viewed 10.3.16]

34 Does insulation paint make any difference? (2011) *The Telegraph* www.telegraph.co.uk/finance/property/advice/8688869/Does-insulation-paint-make-any-difference.html [viewed 10.3.16]

35 Markus, T. and Nelson, I. (1985) *An Investigation of Condensation Dampness at Darnley*, University of Strathclyde, January

36 Menon, R. and Porteous, C. (2011) *Design Guide: Healthy Low Energy Home Laundering*, MEARU (Mackintosh Environmental Architecture Research Unit), The Glasgow School of Art

37 Nearly half of us live in draughty, damp or mouldy homes: Problem found to be worse among rental properties, Mail Online Newswww.dailymail.co.uk/news/article-2718558/Nearly-half-live-draughty-damp-mouldy-homes-Problem-worse-rental-properties.html [viewed 26.2.16]

38 Schofield, C. (2004) Research into mould and the implications for chartered surveyors: Professional briefing for chartered surveyors, 2004/200/RICS Built Environment/40235/

39 Vesper, S.J., Wymer, L.J., Meklin, T., Varma, M., Stott, R., Richardson, M. and Haugland, R.A. (2005) Comparison of populations of mould species in homes in the UK and USA using mould-specific quantitative, *PCR Letters in Applied Microbiology* 41, 367–373

40 Vesper, S., Wymer, L., Kennedy, S. and Grimsley, L. (2013) Decreased Pulmonary Function Measured in Children Exposed to High Environmental Relative Moldiness Index Homes. *The Open Respiratory Medicine Journal* 7 (1), 83–86

41 Hunter, C.A. and Lea, R.G. (1994) The airborne fungal population of representative British homes. *Air Quality Monographs 2*. Health Implications of Fungi in Indoor Environments

42 Sharpe, R.A., Bearman, N., Thornton, C.R., Husk, K. and Osborne, N.J. (2015) Indoor fungal diversity and asthma: a meta-analysis and systematic review of risk factors. *Allergy Clin Immunology* 135 (1), 110–122

43 Chasseur, C., Bladt, S. and Wanlin, M. (2015) Index of indoor airborne fungal spores pollution in Brussels habitat, Healthy Buildings Europe Conference, Eindhoven

44 Burge, H.A. (2001) Fungi: toxic killers or unavoidable nuisances? *Annals of Allergy, Asthma and Immunology* 87, Supplement 3, 52–56

45 Kuhn, D.M. and Ghannoum, M.A. (2003) Indoor mold, toxigenic fungi, and Stachybotrys chartarum: infectious disease perspective, *Clinical Microbiology Reviews* 16, 144–172

46 WHO (2009) *WHO guidelines for IAQ: dampness and mould*, WHO, Geneva

47 Burge (2001) op cit.

48 Terr, A.I. (2001) Stachybotrys: relevance to human disease. *Annals of Allergy, Asthma and Immunology* 87, Supplement 3, 57–63
49 Kuhn and Ghannoum (2003) op cit.
50 Wilkins, K., Larsen, K. and Simkus, M. (2000) Volatile metabolites from mold growth on building materials and synthetic media. *Chemosphere* 41 (3), 437–446
51 WHO (2009) op cit.
52 Hetreed, J. (2012) *The Damp House*, Crowood Press
53 Menon, R. and Porteous, C. (2011) *Design Guide: Healthy Low Energy Home Laundering*, MEARU
54 Liverpool Healthy Homes: Controlling Condensation and Mould (undated) http://liverpool.gov.uk/council/strategies-plans-and-policies/housing/healthy-homes-programme/healthy-homes-what-we-do/ [viewed 26.2.16]
55 Johannson, P. *et al.* (2005) Microbiological growth on building materials critical moisture level. State of the art. 11 SP, Swedish National Testing and Research Institute
56 Hens, H. (2007) *Building Physics: Heat Air and Moisture*, Ernst and Sohn
57 Seppänen, O. and Kurnitski, J. (2009) *WHO Guidelines for IAQ Dampness and Mould* (Chapter 3)
58 Johannson, *et al.* (2005) op cit.
59 Alberg, O. (2011) Moisture and mould problems a threat against sustainable buildings. The BETSI study in Sweden, The National Board of Housing Building and Planning, Sweden
60 Johansson, P. (2012) *Critical Moisture Conditions for Mould Growth on Building Materials 2012*, Rapport TVBH-3051 Lund Avdelningen för Byggnadsfysik, LTH
61 Friman, J. (2010) *Comparative Study on Mould Growth on Plaster Boards treated with Biocides*, Plant Ecology Department of Plant and Environmental Science, Göteborg University
62 Tarmac Building Products Limelite CPD Presentation (2013) Version 2.0
63 Tarmac Building Products Limelite CPD Presentation (2013) op cit.
64 Mould & Mildew Control, Kingsfisher www.kingfisheruk.com/section_mould_control [viewed 26.2.16]
65 Technical Q&A 18: Limewash www.spab.org.uk/advice/technical-qas/technical-qa-18-limewash/ [viewed 26.2.16]
66 Maintenance Matters Limewash, CADW http://cadw.gov.wales/docs/cadw/publications/Maintenance_Matters_Limewash_EN.pdf [viewed 26.2.16]
67 SD70 Safety Data Sheet, Natural Hydraulic Lime Range www.singletonbirch.co.uk/media/downloads/naturalhydrauliclimerange.pdf [viewed 26.2.16]
68 The Building Lime Forum https://buildinglimesforum.org.uk/ [viewed 26.2.16]
69 Holmes, S. and Wingate, M. (2002) *Building with Lime*, ITDG Publishing
70 Bryce, K. and Weisman, A. (2015) *Clay and Lime Renders Plaster and Paints*, Green Books
71 Guelberth, C.R. and Chiras, D. (2002) *The Natural Plaster Book: Earth, Lime, and Gypsum Plasters for Natural Homes*, New Society Publishers Canada
72 Harrison, H.W., Trotman, P.M. and Saunders, G.K. (2009) *Roofs and Roofing Performance, Diagnosis, Maintenance, Repair and the Avoidance of Defects*, Third edition, BRE Press
73 Stramit StrawBoard www.stramit.co.uk/ [viewed 26.2.16]
74 Seeking views on moisture risk assessment and guidance, Department of Energy and Climate Change, 15 April 201
75 May, N. and Saunders, C. (2014) *Moisture Risk Assessment and Guidance Technical Appendix*. DECC
76 STBA (2012) *Responsible Retrofit of Traditional Buildings*, STBA
77 Clarke, C. (2014) *STBA / SPAB Technical Panel Rye C.* The SPAB Building Performance Survey: 2014 Energy Efficiency Research Update Conference 2014
78 Historic Scotland (2012) Internal wall insulation to six tenement flats, Refurbishment Case, Study 4, Sword Street, Glasgow
79 Hay, S. (2013) Green Deal Energy Company Obligation and Traditional Buildings www.historic-scotland.gov.uk/technicalpaper17.pdf [viewed 26.2.16]

80 Roels, S. and Water, C.J. (2001) Vapour permeability and sorption isotherm of coated gypsum board, *Journal of Building Physics* 24 (3), 183–210

81 Fraunhofer Institute Bauphysik (2004) Moisture buffering effects of interior linings made from wood or wood based products, IBP Report HTB-04/2004/e

82 Rode, C. (2005) NORDTEST Moisture Buffering of Building Materials, Department of Civil Engineering Technical University of Denmark, Report BYG·DTU R-126

83 Salonvaara, M., Ojanen, T., Holm, A., Künzel, H.M. and Karagiozis, A.N. (2004) *Moisture Buffering Effects on Indoor Air Quality*, Experimental and Simulation Results, Ashrae

84 Padfield, T. and Aasbjerg Jensen, L. (2010) Humidity buffer capacity of unfired brick and other building materials www.conservationphysics.org/wallbuff/buffer_performance.pdf [viewed 26.2.16]

85 Morton, T., Stevenson, F., Taylor, B. and Charlton Smith, N. (2005) Low Cost Earth Brick Construction 2 Kirk Park, Dalguise: Monitoring & Evaluation, ARC-Architects Low-Cost-Earth-Masonry-Monitoring-Evaluation-Report-2005.pdf www.arc-architects.com/downloads/ [viewed 26.2.16]

86 Morton, T. (2008) *Earth Masonry Design and Construction Guidelines*, BRE Press

87 Padfield, T. (1999) *Humidity buffering of interior spaces by porous, absorbent insulation*, Department of structural Engineering and materials, Technical University of Denmark

88 Mumovic, D., Davies, M., Ridley, I., Oreszczyn, T., Bell, M., Smith, M. and Miles-Shenton, D. (2005) An evaluation of the hygrothermal performance of 'standard' and 'as built' construction details using steady state and transient modeling. Leeds Becket University

89 BSI (2002) Hygrothermal performance of building components and building elements: Internal surface temperature to avoid critical surface humidity and interstitial condensation. Calculation methods, BSI, London, BS EN ISO 13788:2002

90 BSI (2002) BS 5250: 2002 Code of practice for control of condensation in buildings, BSI

91 BS 4016 1997 Specification for flexible building membranes breather type

92 BS 7374 1990 Methods of test for water vapour transmission resistance of board materials

93 Mundt-Petersent, O.S. and Haredrup, L. (2015) Predicting Hygrothermal Performance in Cold Roofs Using a 1D Transient Heat and Moisture Calculation Tool, *Journal of Building and Environment* 90, 215–231

94 Oreszczyn, T., Mumovic, D., Davies, M., Ridley, I., Bell, M., Smith, M. and Miles-Shenton, D. (2011) Condensation risk – impact of improvements to Part L and robust details on Part C, Final report: BD2414 Department for Communities and Local Government

95 www.astm.org [viewed 26.2.16]

96 BuildDesk Software www.builddesk.co.uk/software/ [viewed 26.2.16]

97 IES-VE User Forum http://forums.iesve.com/viewtopic.php?t=1818 [viewed 29.2.16]

98 Butt, T.K. (2005) Water Resistance and Vapor Permeance of Weather Resistive Barriers, *Journal of ASTM International* 2 (10), Paper ID JAI12495

99 Prangnell, R.D. (1971) The water vapour resistivity of building materials a literature survey, *Matériaux et Construction* 4 (6), 399–405

100 Breathability & Moisture Vapour Permeability, Keim Mineral Paints http://keimpaints.co.uk/fileadmin/uk/pdf/Breathability___Moisture_Vapour_Permeability.pdf [viewed 29.2.16]

101 Water Vapor Diffusion, WUFI www.wufi-wiki.com/mediawiki/index.php5/Details: Water VaporDiffusion [viewed 29.2.16]

102 NHBC (2012) Technical Extra 6

103 The Intelligent Member Association www.imaroofer.com/ [viewed 29.2.16]

104 May. N., Why we can't rely on current moisture calculations in traditional buildings, Natural Building Technologies www.natural-building.co.uk/sites/natural-building.co.uk/files/pdf/47/breathability-and-getting-accurate-moisture-calculations.pdf [viewed 29.2.16]

105 Quotes: Dew Point Analysis vs. Hygrothermal Simulation, The Building Enclosure https://builtenv.wordpress.com/2014/03/09/quotes-dew-point-analysis-vs-hygrothermal-simulation/ [viewed 29.2.16]

106 Kingspan Insulation Leominster (2009) *Breathability – A White Paper: A study into the impact of breathability on condensation, mould growth, dust mite populations and health*, Kingspan
107 Mulligan, H. and Brown, A. (2008) *Breathability and the Building Envelope*, Cambridge Architectural Research Prepared for Kingspan Insulation Ltd.
108 May, N. (2009) Breathability Matters: Why the Kingspan White Paper is misleading, unpublished
109 A survey of the breathable building structure concept – Effect Insulation Material Research report VTT S 08448-11, November 25 2011

8 Ventilation and a critique of Passiv Haus

This book is primarily about construction materials and their impact on IAQ and health. However it is necessary to address the issue of ventilation, understanding the importance of good ventilation to ensuring good IAQ. Furthermore there is increasing pressure, as buildings become more and more energy efficient and airtight, to assume that mechanical ventilation is necessary and essential to healthier buildings. While this is normal in deep plan offices and other commercial buildings, it is increasingly assumed that mechanical ventilation is also necessary for houses, schools and smaller offices and that this may become compulsory under building regulations in the future. While it is essential to have good ventilation, it should not be assumed that this requires mechanical air handling and natural and manual ventilation can often be sufficient.

Increased airtightness

Measures such as trickle vents in windows and relying on occupants to open windows are seen by many experts as insufficient to ensure good ventilation when buildings are more airtight. Extreme energy efficiency, as advocated by Passiv Haus (PH), is dependent on mechanical ventilation and heat recovery systems, however others still believe that natural and passive ventilation is all that is required for smaller buildings. Opening windows, in the relatively temperate climate of the UK and much of Europe, is still a simple and adequate way to let fresh air into buildings. As discussed in Chapter 7, the use of hygroscopic materials to manage moisture may also reduce demands on ventilation systems.

It is not possible to resolve the debate about mechanical versus natural ventilation in this book but it is important to issue a warning that claims that MVHR systems are essential to resolve IAQ problems may not be correct. An increasing dependence on mechanical ventilation may exacerbate, rather than resolve, IAQ problems. Relatively low levels of air change may be sufficient to provide enough air to comply with building regulations and for occupants to breathe but this may not be enough to remove concentrations of VOCs or hazardous chemicals and particulates. Purge ventilation, particularly in the early weeks of a new building, is necessary to reduce emissions from new materials, but how many householders go out to work for the day and leave all the windows in their new house wide open?

Some of the literature about IAQ tends to focus almost entirely on ventilation and air conditioning, while ignoring construction materials and the sources of pollutants. Rather than removing the pollution at source or preventing it happening in the first place, it is assumed that indoor air pollutants will not cause health problems if they are removed through mechanical ventilation. Research into this is urgently required and a scientific network investigating the Health Effects of Modern Airtight Construction (HEMAC) has been set up to bring together experts in the field.[1]

Mechanical ventilation

Healthy buildings require a regular exchange of fresh air. As we breathe out CO_2, and as mould and dust particles accumulate, we need to remove these from our buildings. A failure to do so will lead to high levels of mould and greater accumulations of toxic chemicals. Traditionally houses were ventilated through windows being opened, and in older houses windows and doors were draughty, ensuring plenty of 'unwanted' fresh air. Draughts are uncomfortable when the weather is cold, and as buildings become more energy efficient, they have tighter and tighter sealed windows and doors. Occupants need to manage airtight buildings more effectively by opening windows and purging air when possible. Trickle vents in windows were required by building regulations but are rarely effective as a means of ventilation. Many people seal them up as they think they cause draughts. Tens of thousands of houses have replaced their old timber or metal windows with uPVC but in thousands of cases uPVC replacement windows have been installed without trickle vents. Wet rooms such as kitchens and bathrooms have inefficient, intermittent mechanical extract fans, which can be noisy and expensive to run and can get blocked as householders do not realise they have to be cleaned out from time to time.

Powerful industrial and commercial lobby groups have been lobbying in recent years to make mechanical ventilation compulsory in housing. Advocates of extreme energy efficiency such as PH claim that mechanical ventilation ensures better air quality. 'The ventilation systems used in Passive Houses provide unparalleled IAQ through the use of a high quality, F7 filter at the suction point'.[2] PH advocates, such as Justin Bere Architects, say that research proves that PH ensures better air quality than in conventional houses. But as will be shown below this is not always the case.

> Air quality tests carried out by Derrick Crump of Cranfield University for Bere architects, and funded by the TSB, may provide some early indications that a combination of air-tight window seals and heat recovery ventilation systems may significantly reduce levels of harmful particulates … VOC levels were also found to be low in all three certified Passive Houses tested, an important factor in air quality.[3]

There are a number of key questions here;

1 Does mechanical ventilation ensure that pollutants inside buildings from construction materials, etc. are removed?
2 Are there dangers that air-handling systems, introducing air from outside, are injecting more pollutants into buildings or can they reduce internal air pollution?
3 Is the air from mechanical systems clean and fresh air or is it contaminated by accumulations in filters and emissions from the materials used to make the ducts and heat exchangers?
4 Is natural and passive ventilation an effective alternative option?

These are complicated questions and insufficient research has been carried out to resolve them, as academics and engineers are generally convinced that mechanical ventilation is the only answer. Publications advocating mechanical ventilation and PH largely originate from practitioners or advocates of PH or from commercial companies selling the equipment. Independent academic evaluation of PH and MVHR systems, by contrast, is largely critical of IAQ problems associated with PH and MVHR. 'Complaints about the indoor environment indicate that certain health complaints are *more prevalent with heat recovery ventilation systems* ... Smaller and well-occupied passive houses are more at risk'.[4]

Hasselaar cites research by Ginkel showing significantly high levels of perceived health problems in mechanically ventilated houses when compared with conventional housing.[5] Despite this evidence, some continue to promote the idea of mechanical ventilation having health benefits. A campaign begun in 2015 puts forward the case for all newly built houses to display a 'Healthy Home Mark'. However on investigation it becomes clear that a Healthy Home Mark would not necessarily provide evidence of good IAQ but simply that mechanical ventilation has been installed.

> to help protect and improve IAQ, and the health of future generations. The campaign calls for support of a 'Healthy Home Mark' in all new homes to allow home buyers to identify that the dwelling has been fitted with effectively installed continuous mechanical ventilation, which delivers healthy indoor air.[6,7]

The Healthy Home Mark is in fact an advertising promotion for mechanical ventilation systems by BEAMA, a trade association for the mechanical ventilation and electrical infrastructure products and transmission systems industry.[8] BEAMA is both a trade body and a lobby group with offices near the Houses of Parliament. It commissioned a YouGov consumer survey reported in *Heating, Ventilating and Plumbing* magazine, which found that 58 per cent of respondents in a survey of households suffered from mould or condensation in their homes, with 19 per cent of these residents already having suffered from a respiratory or dermatological condition. The article also reports on a study by Waverton Analytics, which found that 91 per cent of 120 UK homes had excessive levels of TVOCs.[9] BEAMA see the solution to these problems as requiring mechanical ventilation and bases this on a report by Professor Awbi of Reading University who argues that MVHR is the solution to IAQ problems. There is no doubt that good ventilation is essential for healthy buildings, but Awbi discounts the importance of dealing with hazardous materials at source. Awbi and BEAMA appear to accept that buildings will produce volatile emissions and then argue that this can be mitigated through *forced* ventilation.[10]

Awbi argues that 'the only way to control indoor pollutant levels (including CO_2) is by proper ventilation. Indoor pollutants are not just produced from building materials and furnishing but *mainly* from household activities, such as cooking, cleaning, cosmetics, etc.'[11] While cooking, cleaning and cosmetics can contribute to indoor pollutants and some air fresheners can be a serious source of VOCs, this is not an argument for ignoring the other pollutants from building materials. Awbi's report provides a cogent reminder of the problems of airtight dwellings. He quotes the NHBC, claiming that 'a lack of adequate ventilation ... could result in a build up of pollutants released by furnishings, building insulation materials, etc.' Awbi refers to several other studies with similar conclusions and states 'it was not established whether (pollutant) concentrations were higher because of stronger sources of (these) compounds or whether lower ventilation rates occurred'. Having made clear that there is a lack of data on the causes of pollutant emission problems he states that these studies 'provide substantial evidence on the link between low ventilation rates and the degradation of indoor air quality', even though he himself states that the evidence is not there![12] As Awbi's study was commissioned by BEAMA, it should not be seen as independent evidence and campaigns such as 'My Health My Home', set up by BEAMA, are vehicles for promoting commercial mechanical ventilation equipment.[13]

Ventilation manufacturers including Vent Axia and Envirovent have issued press releases referring to a so-called 'Pan-European' study carried out by the Finnish National Institute for Health and Welfare, which, it is claimed, demonstrates a link between indoor air pollutants and cardiovascular disease, and that changes in the ventilation of homes, would lead to a 38 per cent reduction in this disease.[14]

While the industry will be keen to promote mechanical ventilation, it is necessary to make clear that mechanical ventilation may not be able to remove persistent pollutants from construction materials in a building. Ventilation performance is a complicated science dealing with air changes and air movement. High-powered extract fans can replace air in a room very quickly but this is not how most ventilation in homes works. Stale air can be displaced by incoming air but in most cases it is mixed with existing air and according to Bluyssen 'While from an air quality aspect, displacement approaches are generally preferred, very precise operating conditions are normally needed and their operation can be impaired by occupant activities, temperature fluctuations and door openings etc'.[15]

Bluyssen points out that air ducts are frequently left lying around on sites during construction and when installed contain ready-made pollutants from dirt and dust and microorganisms. This is something I have observed myself in projects where MVHR systems are being installed. In one project, scraps of fibreglass insulation could be seen all over the interior of the buildings and inside ducts, whilst ductwork was being installed and nothing was being done to clean these out and protect against contamination.

VOCs, other pollutants and mould growth are adsorbed onto surfaces and absorbed into fabrics in buildings. Mechanical ventilation systems may not entirely deal with this problem. Good regular ventilation can inhibit the development of mould growth for instance but may not always be able to get rid of it once it is established. Good ventilation may dilute the levels of indoor air pollution but will not remove the source of emissions.

There is little doubt that purge ventilation is effective in dissipating some pollutants, especially in the early days of a new building, or after building work or painting has been done. However it is not always practicable to leave all the doors and windows open because of security concerns. Opening windows in the morning and evening may not be enough. Cooker hoods and extractor fans in bathrooms can have similar limitations. It is for this reason that well designed ventilation is necessary and it is not suggested here that mechanical ventilation be ruled out but sources of hazardous emissions must also be removed or avoided.

Passiv Haus – is it healthy?

A major difference in approach to the design of buildings has occurred as a result of the popularity of the 'passive house' concept. *Passive* house is mis-named as it implies that the house works in a passive way, whereas, due to its dependence on mechanical ventilation, it relies on active electrically powered systems. The use of the word passive should not be confused with the idea of passive solar gain, which was an early and popular concept in ecological design, where windows were concentrated on the south facing elevations and thermal mass was used to store warmth from the sun. Passive design[16] has become less popular when it was recognised that over-glazed buildings can overheat. Using solar gain to heat a building has some benefit but has to be managed with care.

The PH concept advocates extremely high levels of airtightness and high levels of insulation in an effort to achieve very high standards of energy efficiency. It has been claimed that only a heated towel rail is required to warm the house.[17]

The Passiv Haus Planning Package (PHPP), a modelling technique to predict performance, has become very popular with many architects and builders and some public authorities have tried to make the PH standard compulsory in an effort to meet carbon reduction targets. Due to the high levels of airtightness, PH requires the installation of a mechanical ventilation system. It is claimed that the energy required to run the MVHR system is very low. It is also claimed that the heat recovered through such a system will reduce the need for heating systems.

Most certified PH projects are built with conventional masonry or timber-frame construction using petrochemical-based insulation materials and it has not been possible to find anything in the PH standards about safeguarding occupants from indoor pollution emissions such as formaldehyde, VOCs and other hazardous materials. There are some examples of PH projects that have been built using ecological and natural materials, but under the PH standard, IAQ is entirely dependent on MVHR. 'Virtually all construction methods can be successfully utilised for Passivhaus design. Masonry (cavity wall and monolithic), timber frame, off-site prefabricated elements, insulated concrete formwork; steel, strawbale and many hybrid constructions have been successfully used in Passivhaus buildings'.[18]

> A Passivhaus … is required to achieve sufficient indoor air quality conditions – without the need for additional recirculation of air … As well as being an energy performance standard Passivhaus also provides excellent IAQ, this is achieved by reducing the air infiltration rates and supplying fresh air which is filtered and post heated by the MVHR unit.[19]

It is possible to divide the building design community into two factions, differentiating those who are convinced that mechanical ventilation is the way forward, even for the smallest of houses, from others who advocate an *'Active House'* concept, where they believe that it is essential for occupants to take responsibility for managing their indoor environment and not be totally dependent on MVHR. A number of businesses and universities have formed the Active House Alliance, partly due to concerns about the dominance of PH.[20]

As referred to above, evidence has emerged over the past few years that the PH approach can be problematic because occupants do not manage the system well or fail to understand it. Having stayed in several PH houses, in every case, the MVHR air input systems were noisy and it is not surprising that occupants, who do not fully understand the concept, tend to switch off the MVHR fans. Tenants in social housing have also been found to be concerned at the cost of running the fans. There is a considerable maintenance burden, with MVHR systems and filters that are not changed regularly and can get clogged with dirt, reducing the ingress of fresh air.

Bjorn Berge, a leading authority on ecological building has warned of the dangers of becoming 'dependent on the machine'.

> In building policy it is not the aim to open up to simple approaches. It is thus quite logical that natural ventilation is rejected in favor of more complex and highly mechanized ventilation strategies … the passive house will … eventually have to be killed off … And so rigid and specialized as these building types are now about to be, there will be little opportunity for reconstruction and adaptation. There will therefore be a lot of demolition.[21]

Berge's critique may seem to be anti-technology but has a more fundamental philosophical principle at its core in that as society becomes increasingly dependent on specialised mechanical components, ordinary people can no longer manage even a simple environment such as the home. Hasselaar echoes Berge's concerns warning of a 'technology driven promotion and the strict application of the concept creates increased risk of *design dictatorship*'.

> The promotion of the idea that the user must 'be educated' and that SMART solutions are needed to prevent wrong behavior by the occupants, shows the need for participative design methods and for more flexibility in control functions for the users. The major recommendations are: design of hybrid ventilation system, reduction of the noise level of HRV ventilation, frequent cleaning of the components of the air inlet system, good sun shading and innovation for flexible heating sources.[22]

Hens has also warned of the dangers in a case study of a failed PH project in Belgium, and advocates moisture-tolerant building fabric.

> Passive Houses: what may happen when energy efficiency becomes the only paradigm? A healthy indoor environment is a prime prerequisite and a moisture tolerant building fabric is a basic durability-related performance requirement. Both have been largely overlooked in the passive dwelling case discussed.[23]

In Belgium a number of PH projects have been built with underground heat recovery systems, which led to such systems being scrapped, and in one case, a house being declared unfit for human habitation. There are also concerns about elevated radon levels in PH houses. In one study, a radon level of 700 Becquerels was found in PH house, as well as elevated CO_2 levels.[24] McGill *et al.* carried out a study of houses built to the PH standard. Occupants found the houses were too warm in the summer, though one was using heating during most seasons. CO_2 levels in one of the houses exceeded 1,000 ppm during both summer and winter, raising concerns about poor IAQ.

> The results from the occupant interviews suggest significant issues regarding maintenance and use of the Mechanical Ventilation with Heat Recovery (MVHR) system. For instance, inadequate knowledge of the boost mode function and the importance of 'balancing' the MVHR system, adjustment of the supply and extract vents, on-going system faults, lack of skilled service engineers, lack of filter replacements, uncertainties over responsibility of maintenance requirements, problems with noise on higher settings, draughts and problems with thermal comfort were all identified through the interview process.[25]

The majority of PH projects use petrochemical and synthetic materials for insulation and PH advocates claim that mechanical ventilations systems are adequate to maintain good IAQ, despite possible emissions from these materials. Bere even claims that PH is 'health enhancing', despite the growing number of studies identifying serious problems or poor air quality associated with PH. 'This is yet further solid evidence of the health-enhancing benefits of a Passive House, and firm evidence of the health-enhancing potential of properly designed and installed heat recovery ventilation systems'.[26]

Studies dating back to 2008 have identified health and air quality problems in PH and other highly energy efficient house projects according to both the NHBC and the UK's Zero

Carbon Hub. Most of these problems are associated with MVHR systems.

> In general, the air volumes tend to fall back because of clogging of filters and dirt on fan blades, asked for over-capacity in the design stage. The two inspected cases reveal poor attention to inspection and cleaning of ducts. Pollution from deposited dirt in the system can become a health risk after some years.[27]

> Given the weight of evidence suggesting that all is not well, the potential for poor IAQ and health issues, and the potential for any energy/CO_2 benefit not to be realised, the Task Group considers that there is a need for further monitoring of MVHR systems in use post-occupancy.[28]

MVHR systems will not necessarily deal with emissions from hazardous materials, as has been shown in several studies. 'Daily peaks in formaldehyde concentration were reported in two houses; one of the graphs shows a concentration of 1.6 ppm in one case which, if accurate, considerably exceeds the WHO guideline value for formaldehyde (WHO, 2000)'.[29] A Swedish study found higher levels of TVOCs in newly built PH houses, though lower levels of formaldehyde when compared with conventional houses.[30] Nash warns of the dangers of 'living in a plastic bag' and the risks to IAQ and dependence on MVHR systems and recommends that developers should build with lower emissions materials.[31]

Adoption of the PH standard has already become a political battleground. In the Republic of Ireland, for instance, the Government Minister for the Environment Alan Kelly instructed some city councils in Ireland that they were not to impose the PH standard.[32] Dun Laoghaire and Rathdown Council has attempted to make PH mandatory for all new buildings through its planning policy, but central government in Ireland has been investigating various forms of legal challenge to this. Forcing the adoption of PH standards may be seen as a breach of EU procurement rules as PH is a private commercial operation and not a generic standard. There have also been rumours that achieving PH standards for new builds could become compulsory in the UK, though this is unlikely under the present UK government that has scrapped zero CO_2 targets.

Despite extravagant claims about the energy performance of PH buildings, the certification of PH projects is not carried out independently but by exponents of the PH approach. Independent research in the Netherlands found houses renovated to PH standards had a tendency to overheat, though their computer simulation tests showed that the MVHR system in the houses could be adjusted to reduce this problem. However the researchers also recommended night-time cooling measures such as opening windows and increased shading. 'The increasing number of highly insulated and air tight buildings leads to the concern of indoor environment overheating and related comfort and health issues. This can already happen in a temperate climate as found in the Netherlands'.[33]

Sassi found convincing evidence that a naturally ventilated flat in Cardiff used less energy than predicted by the PHPP, with the naturally ventilated flat using less energy than the one done to PH standards. Her study considered:

> the comfort, air quality and energy impacts of MVHR versus natural ventilation and reviews the post-occupancy monitoring data of two flats in Cardiff designed to Passivhaus standards, one of which had been operated as a naturally ventilated building rather than with MVHR. The energy consumption of this free-running flat was significantly lower (36 kWh primary energy/mÇa) than the Passivhaus Planning Package modeling

> had predicted (93 kWh primary energy/mÇa) with no adverse effects on occupant comfort, air quality or excessive humidity, and advantages of lower capital cost and maintenance. The paper concludes that in climates with mild winters and cool summers the use of MVHR could be omitted without compromising comfort levels and achieving at least equivalent energy savings resulting from adopting the Passivhaus model and at a lower capital cost. This suggests the potential for a naturally ventilated, ultra-low energy model with lower capital investment requirements and lower disruption when applied to retrofit that would facilitate its mainstream adoption.[34]

The majority of PH projects are insulated with non-breathable petrochemical-based insulation materials and tightly wrapped with plastic membranes. Thus it is not surprising that PH has attracted the support of chemical and synthetic insulation companies.

> The thermal insulation requirements of Passiv haus demand that roof, wall and floor U–values are equal to or less than 0.15 W/m2.K. Kingspan Insulation offers amongst the thinnest commonly available insulation, meaning that space taken to achieve the stringent U–value is kept to a minimum. Over 30,000 buildings have been constructed in accordance with Passiv haus principles, with approximately 70 projects certified or in progress in the UK.[35]

PH certified and other projects, claiming to follow PH principles, have been built using natural and ecological materials, though most still use MVHR systems. A German company, Bio-Solar-Haus, is highly critical of standard PH 'technology house' approaches and claims to achieve equivalent energy efficiency standards while avoiding the use of plastic moisture barriers and 'problematic ventilation systems'.[36]

Natural, passive and purge ventilation

Given the growing evidence of concerns about MVHR systems and their ability to safeguard good IAQ, it is important to assess whether natural ventilation measures can ensure good air quality. 'Fresh air' is a complicated concept because in cities, opening the windows will let in polluted air from traffic and other sources. Even in the countryside, pesticides and other agricultural chemicals may be allowed into the house. However we all need to breathe clean air and outside is normally the best place to find it.

Even though numerous authorities state that poor IAQ is much more dangerous than external air pollution, many governments have chosen to focus on external air problems at the expense of research and policy on indoor air. Living in an artificially ventilated environment for much of the time is not good for health, but many offices, schools and healthcare buildings are now dependent on mechanical air handling. Opening windows is less and less common. However medical research has shown that opening windows is much better for health than mechanical ventilation. Jopson argues that 'regularly cleaned hospital rooms still have fewer bacteria overall … but if the windows are opened between the cleaning, the bacteria that are present will be less likely to cause a disease or an infection'.[37]

Gilbert, a microbiologist at the Argonne National Laboratory Biosciences Division, advocates opening hospital windows to stem spread of infections, something that Florence Nightingale believed in many years before![38] Gilbert's views reflect work by Kembel *et al.*, examining microbial samples in buildings and this study also refers to natural ventilation from open windows.

> the effect of ventilating rooms with outdoor air entering directly through open windows: Our findings suggest that it is worthwhile to explore the effects of natural ventilation on microbial communities more rigorously using modern molecular tools, as we found that the abundance of potentially pathogenic bacteria was not higher in window-ventilated patient rooms than in mechanically ventilated rooms.[39]

The active house concept

> [Active House] supports the vision of buildings that create healthier and more comfortable lives for their residents without impacting negatively on the climate and environment – thus moving us towards a cleaner, healthier and safer world. Materials used have a positive impact on comfort and indoor climate ... through the use of low emitting materials.
>
> The Active House measures a lot of things that the Passive House doesn't, such as light and view, indoor air quality, natural ventilation and daylighting, all of which can be improved by, you guessed it, more opening windows and skylights.[40]

There is an assumption, among many architects and engineers, that natural ventilation cannot ensure sufficient air changes or adequate input of clean air even though this has worked for centuries in hot countries, combining thermal mass for cooling and roof vents, shading and courtyards.[41] Modern 'passive stack' ventilation systems are available and have been found to be effective. There are even novel systems, which claim to achieve heat recovery without using mechanical systems, as well as providing natural ventilation.[42]

Another option is to consider humidity sensitive mechanical hybrid ventilation systems. A range of these are now available and are claimed to be much more energy efficient than MVHR and many avoid the need for expensive and complicated ducting systems. Demand controlled ventilation means that the mechanical extract only kicks in when humidity rises. Such hybrid arrangements mean that a ventilation stack can operate in either a completely passive mode or mechanical in the same system. The systems are designed to cope with changing weather and internal conditions.

MVHR systems may be more efficient in terms of heat recovery, if they work efficiently, but a lot more independent research is needed to verify whether this is actually the case. Most data currently available from research bodies such as Fraunhofer have been commissioned by commercial companies. 'The saving is 17.4% when using a demand controlled exhaust ventilation system'.[43] Using completely natural ventilation effectively in a building requires careful thought by designers and considerable attention to microclimates and the siting of buildings. Considerable technical expertise is available to understand natural ventilation. Allard says that 'occupants of naturally ventilated buildings are usually happier with their indoor environments than when they have little control over them', though he admits that the barriers to implementing natural ventilation are numerous and discouraging.[44]

Purge ventilation

Purge ventilation is an important option to maintain good IAQ but it does rely on the building occupants taking responsibility for this. Purge ventilation measures are set out in the UK building regulations as intermittent extraction by opening windows, doors and rooflights and this can be very useful to reduce contaminants. Cooker extractor hoods are also a form of purge ventilation. Careful management of a building can include opening windows and roof lights to provide a good exchange of fresh air without massive heat loss. As it says in the UK building regulations:

> Adequate purge ventilation may be achieved by the use of openable windows and/or external doors of the following sizes:
>
> - A hinged or pivot window that opens 30º or more, or sliding sash window where the height x width of the opening area is at least 1/20th of the floor area of the room.
> - A hinged or pivot window that opens less than 30º, the height x width of the opening area is at least 1/10th of the floor area of the room.
> - An external door (including patio doors) that has an opening area at least 1/20th the floor area of the room.[45]

Despite the criticisms of PH and mechanical ventilation in this chapter, there is no doubt that greater energy efficiency and better ventilation in buildings are essential and a good thing. What is important is that we have the best and most reliable forms of ventilation available, both natural and mechanical, and that energy efficiency is not achieved at the expense of good air quality and lower impact materials. The myopic use of petrochemical materials to achieve energy savings is mistaken, as the manufacture of these materials is also contributing to CO_2 emissions and this is not necessarily offset by energy saved in buildings.

Notes

1 https://hemacnetwork.com/about-2/ [viewed 5.5.16]
2 www.passivehouse-international.org/index.php?page_id=80 [viewed 23.2.16]
3 Bere, J. (2013) Air quality in Passive Houses, *RIBA Journal*, 1 November www.ribaj.com/products/air-quality-in-passive-houses [viewed 23.2.16]
4 Hasselaar, E. (2008) Health risk associated with passive houses: An exploration, Indoor Air conference, 17–22 August, Copenhagen, Denmark, Paper ID: 689
5 Ginkel, J.T. van. (2007). Inventarisatie woninggerelateerde gezondheidsklachten in Vathorst, report for Municipality of Amersfoort
6 www.hpmmag.com/ventilation/beama-launch-campaign-for-healthy-home-mark [viewed 1.2.16]
7 www.myhealthmyhome.com/ [viewed 3.3.16]
8 www.beama.org.uk/aboutbeama.html#sthash.DYjQbAmq.dpuf [viewed 3.3.16]
9 http://hvpmag.co.uk/news/fullstory.php/aid/2699/BEAMA_awareness_campaign_highlights_risks_of_poor_indoor_air_quality.html (November 2014) [viewed 3.3.16]
10 Awbi, H.W. (2015) *The Future of Indoor Air Quality in UK Homes and its Impact on Health*, BEAMA
11 Personal emails from Professor Awbi, 28.9.15
12 Awbi, H.W. (2015) op cit.
13 www.myhealthmyhome.com/ [viewed 3.3.15]
14 http://hvpmag.co.uk/news/fullstory.php/aid/3333/Study_finds_link_between_serious_health_problems_and_indoor_air_pollution.htm [viewed 2.3.16]
15 Bluyssen, P.M. (2009) *The Indoor Environment Handbook*, Earthscan and RIBA Publishing
16 Roaf, S. (2001) *Ecohouse 2*, Architectural Press
17 www.passivhaus.org.uk/standard.jsp?id=122 [viewed 1.3.16]

18 Passivhaus primer: Designer's guide, A guide for the design team and local authorities www.passivhaus.org.uk/filelibrary/Primers/KN4430_Passivhaus_Primer_WEB.pdf [viewed 1.3.16]
19 www.passivhaus.org.uk/standard.jsp?id=122 [viewed 1.3.16]
20 www.activehouse.info/ [viewed 1.3.16]
21 Berge, B. (2011) The engine is not responding: A critique of the automatic energy-saving home, *Norway Arkitektur* 1/2011
22 Hasselaar (2008) op cit.
23 www.bsria.co.uk/information-membership/information-centre/library/item/passive-houses-what-may-happen-when-energy-efficiency-becomes-the-only-paradigm/ [viewed 20.2.16]
24 Poffijn, A., Tonet, A.O.B., Dehandschutter, M.R. and Bouland, C. (2012). A pilot study on the air quality in passive houses with particular attention to radon, AARST www.aarst.org/proceedings/2012/07 [viewed 23.3.16]
25 McGill, G., Qina, M. and Oyedele, L. (2014) A case study investigation of IAQ in UK Passivhaus dwellings International Conference on Sustainability in Energy and Buildings 2014, SEB-14, *Energy Procedia* 62, 190–199
26 www.bere.co.uk/blog/healthy-results-in-passive-house-air-quality-tests-comparing-particulates-in-a-passive-house [viewed 3.3.16]
27 Crump, D., Dengel A. and Swainson, M (2009) *Indoor Air Quality in Highly Energy Efficient Homes – A Review*, NHBC Foundation NF18 IHS, BRE Press
28 Zero Carbon Hub (2013) *Mechanical ventilation with heat recovery in new homes final report*. Ventilation and IAQ task group, July
29 Balvers, J., Boxem, G. and de Wit, M. (2008) Indoor air quality in low-energy houses in the Netherlands: does mechanical ventilation provide a healthy indoor environment? *Proceedings of Indoor Air 2008*, Paper 719, 17–22 August, 2008, Copenhagen, Denmark
30 Langer, S., Beko, G., Bloom, E., Widheden, A. and Ekberg, L. (2015) Indoor air quality in passive and conventional new houses in Sweden. *Building and Environment* 93, 92–100
31 Nash, S. (2013) Impact of mechanical ventilation systems on the indoor air quality in highly energy-efficient houses: How it affects human health, EES 2013-169 T, University of Groningen. www.rug.nl/research/portal/files/.../EES-2013-169T_SabrinaNash.pdf [viewed 3.3.16]
32 www.theguardian.com/environment/2015/jun/17/irish-government-in-row-over-passivhaus-eco-building-regulations [viewed 15.2.16]
33 Barbosa, R., Loomans, M.G.L.C., Hensen, J.L.M. and Barták, M. (2015) The impact of increased airflow rates on indoor temperatures of passive house in the Netherlands, *Healthy Buildings Europe Conference Proceedings* May
34 Sassi, P. (201), A Natural Ventilation Alternative to the Passivhaus Standard for a Mild Maritime Climate. *Buildings* 3, 61–78
35 Passivhaus buildings: Case Studies, Kingspan, March 2012 www.kingspaninsulation.co.uk/getattachment/9d9ef282-25c4-442a-9668-2db738d3e90d/Passivhaus-Buildings—Case-Studies.aspx [viewed 3.3.16]
36 www.bio-solar-house.com/bio-solar-house.html [viewed 26.3.16]
37 Jopson, M. (2015) *The Science of Everyday Life,* Michael O'Mara Books
38 www.theguardian.com/science/2012/feb/20/open-hospital-windows-stem-infections [viewed 2.2.16]
39 Kembel, S.W., Jones, E., Kline, J., Northcutt, D., Stenson, J., Womack, A.M., Bohannan, B.J.M., Brown, G.Z. and Green. J.L. (2012) Architectural design influences the diversity and structure of the built environment microbiome, *The ISME Journal* 6, 1469–1479 & 2012 International Society for Microbial Ecology
40 www.treehugger.com/green-architecture/active-house-yet-another-green-building-standard-comes-north-america.html [viewed 15.11.2015]
41 Kubota, T., Hooi, C. and Toe, D. (2015) Application of passive cooling techniques in vernacular houses to modern urban houses: A case study of Malaysia. International Conference Green

Architecture for Sustainable Living and Environment, 29 November 2014, *Social and Behavioural Sciences* 179, 29–39

42 Llanos, M.A.F. and Lipinski, T. (2014) Humidity in dwellings with post occupancy evaluation report. 30 May, Ventive www.ventive.co.uk/wp-content/uploads/2014/06/Ventive_Humidity_Report_29072014.pdf [viewed 10.11.15]

43 www.aereco.fr/wp-content/uploads/2012/10/Fraunhofer-2011.pdf [viewed 10.11.15]

44 Allard, F. (ed.) and Santamouris, M. (1998) *Natural Ventilation in Buildings A Design Handbook*, James and James

45 UK Building Regulations Part F 2010 www.planningportal.gov.uk/buildingregulations/approveddocuments/part [viewed 16.3.16]

9 Emission problems in existing buildings

It is relatively simple, when designing a new building, to apply the precautionary principle by excluding hazardous materials and specifying alternative low-impact solutions. However, this is much more difficult when converting or renovating existing buildings, though many of the same principles apply. DIY, painting and decorating, installing new floors or making minor improvements to a house, office or educational building can introduce hazardous materials and damage IAQ. Existing buildings may already contain hazardous materials such as asbestos, treated timber and insulation materials like fibreglass. In this chapter some of the remedies to tackle existing emissions and problems are discussed. Further measures based on the use of non-hazardous materials are discussed in Chapter 11.

Often the remedies to problems in existing buildings can be quite drastic. Hazardous materials, buried inside the fabric of buildings, are hard to remove, and eliminating emissions from them can be a major task. The Palace of Westminster, the UK Parliament building, for instance, requires major renovation to deal with problems of asbestos, decay, damp and mould, rats and other vermin. To deal with these problems fully will probably involve closing the building altogether: 'Major restoration of the Houses of Parliament without moving MPs and peers out would cost £5.7bn and take 32 years, a report says. If MPs and peers were moved out for six years, the cost would drop to £3.5bn'.[1]

Care homes

Not all problems with existing buildings are as serious as those in the Houses of Parliament, but it is not uncommon to see problems related to IAQ reported in schools and healthcare buildings as well as homes and offices. Care homes can suffer badly from poor IAQ as they are not always ventilated effectively. This is not simply a problem of lack of cleanliness and poor care of the residents, as the physical fabric of the buildings may also be to blame. Older people are more likely to suffer from respiratory conditions and thus having good air quality will help with breathing and general well-being. 'Dirty care home smelled so bad, inspectors picked up offensive odour *outside* the building'.[2] 'Air quality in nursing homes affects lung health of residents'.[3] Even at low levels, IAQ affected respiratory health in elderly people permanently living in nursing homes, with frailty increasing with age.[4,5]

Schools

Schools can be vulnerable to indoor air problems. Children and their teachers will be aware of bad odours, and the impact of hazardous materials on young people is more significant than on older people. There are dozens of reports in the USA media of children feeling ill in school and of bad odours. 'Sickening smell inside Cary Elementary School (Richmond Virginia USA) concerns parents'.[6] One American report claimed that *one in five* American schools had indoor air problems, with VOC concentrations exceeding 1,000 $\mu g/m^3$ where ventilation was poor and that CO_2 levels over 1,000 ppm were also commonplace.[7] So widespread is the problem in the USA that parents have formed a Healthy Schools Network, which provides guidance to parents.[8]

> Indoor levels of pollutants may be two to five times higher than the levels of pollutants found outdoors, and this figure can sometimes rise to more than 100 times higher than outdoor levels within schools and other large facilities. In many of today's schools, students are crowded closely together, with class sizes being far larger than they were just a few decades ago. This situation has gotten so extreme that the typical school has approximately four times as many occupants as office buildings for the same amount of floor space, causing bacteria and other contaminants to build up.[9]

Official bodies in the USA, such as Vermont Department of Health, provide guidance on IAQ.[10] Given that the standard of American school buildings is no worse than those in the UK, it is puzzling why similar reports and guidance are largely absent in the UK and Ireland. Examples in Germany and Scandinavia are also not hard to find. Perhaps schools in the UK do not suffer from such problems, but this seems very unlikely. One study of IAQ in schools in the UK relies almost entirely in literature from USA, Germany and Scandinavia, but does provide some useful information on CO_2 levels that exceed safe levels in school classrooms. It also provides data on the use of natural ventilation.[11]

Clean Air London has reported that little is known about air quality in schools, though the report seems largely concerned with promoting the sale of air filters in ventilation systems.

> Report reveals few Local Authorities know if schools comply with indoor air quality standards (2012). A new updated report from Clean Air in London reveals that few local authorities know if their buildings use regularly maintained air filters that comply with indoor air quality standard EN 13779, particularly schools.[12]

A study carried out by the BRE for the UK government in 2006 identified serious problems of poor air quality in schools but this did not appear to have prompted any action or policy guidance from government; 21 per cent of the TVOC samples taken exceeded the BRE safe limit of 300 $\mu g/m^3$.

> Levels of volatile organic compounds (VOCs) were measured in each classroom for a 45 minute period each afternoon (concurrently with ventilation measurements). Levels of total volatile organic compounds (TVOCs) were compared against a proposed guideline value of 300 $\mu g\ m^{-3}$. It was found that 21% of samples exceeded this limit and these high values originated from four schools. The highest level observed was approximately 700 $\mu g\ m^{-3}$.[13]

Studies that have been made in UK schools tend to focus on CO_2 levels,[14] while VOC emissions are ignored. Stories of children falling ill in large groups are invariably put down to mass hysteria and investigation of IAQ problems is not necessarily carried out.[15] CO_2 detectors are sometimes placed in classrooms in new school buildings and when the alarm on the detector sounds after the room has filled with pupils, and CO_2 levels are over 1,000 ppm, the normal action taken by the teacher is not to open windows but to disable the alarm!

Dealing with emissions of hazardous materials

If IAQ tests have been carried in out schools, these may reveal a range of hazardous and carcinogenic emissions, with formaldehyde being the most common and dangerous. Sometimes problems are identified by building occupants who complain of unpleasant odours, but not all dangerous chemicals can be detected by the nose alone and IAQ tests should be carried out to determine what chemicals are present. Once the problem has been detected and diagnosed, there remains the issue of how to deal with it. If the hazardous chemicals are an integral part of the materials used in the building construction and are trapped in the fabric of walls, floors and other parts of the building, then it may not practical to remove the pollutant sources without removing these materials, which may not be easy to do without causing significant disruption.

In some cases the drastic solution of ripping out materials has to be adopted and this can involve high costs to either the building owner or the construction company that ignored

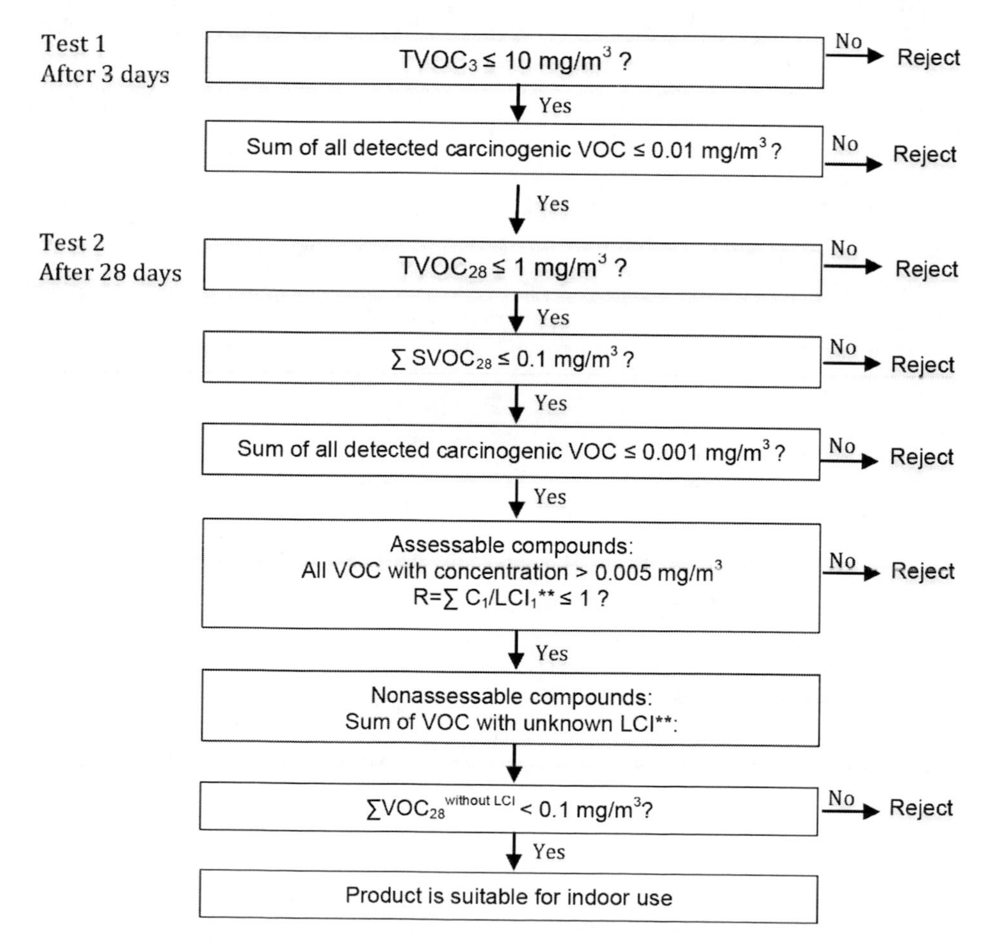

Figure 9.1 AgBB flow chart for evaluation of VOC emissions

IAQ requirements. As it is highly unusual in the UK for materials to be removed due to hazardous emissions it is necessary to look at Germany for examples where this has occurred. 'School in Nuremberg that had been refurbished with rubber flooring. Occupants complained of annoying smells in 2008. Indoor air quality testing revealed naphthalene concentrations from the glue used to stick down the flooring at 50 $\mu g/m^3$ whereas regulations limited naphthalene to 7–17 $\mu g/m^3$'.[16]

Daumling has reported on cases where flooring has been taken up because of the emissions from the glues used. Germany has tried to impose very rigorous standards for emissions in relation to flooring but has had difficulty applying these. The AgBB (Committee for the Health-related Evaluation of Building Products) has found cases where emissions were three times higher than allowed under the German standards. Naphthalene was one of the most serious chemicals but problems were found with other VOCs. Daumling raises the question of who pays for the removal of materials when issues come to court. The AgBB has laid down procedures to be followed when assessing emissions (Figure 9.1) and following these can involve quite a lot of work.[17]

Absorbing formaldehyde and other VOCs

There are measures that can be taken to mitigate problems of VOC emissions by installing materials that will absorb or adsorb toxic chemicals. Some are commercially available, using wool products such as insulation or carpeting. Another option is to use a carbon-based fabric specifically formulated to absorb moisture and chemicals.

Sheep's wool insulation

Wool is a natural material that has been used for clothing, carpeting and furnishings for thousands of years. Despite increased use of synthetic fabrics, wool is still used widely today and is preferred by many people to synthetic alternatives. The fire safety aspects of wool are generally understood as it is difficult to set fire to wool, while synthetic fabrics are often treated with flame retardants. Many clothing products are blends of wool and synthetic fibres and so remain a fire risk, even if the majority of fibre is wool. Flame retardants in fabrics and other materials are a source of hazardous emissions and so should be minimised where possible.

Wool is a useful material for thermal and acoustic insulation and its hygroscopic properties, in terms of dealing with moisture, are discussed in Chapter 11. In addition to its hygroscopic properties, natural wool is claimed to have the ability to absorb formaldehyde and other toxic chemicals. Thus if wool insulation and, to some extent, wool fabrics are used in a building, this might reduce the risk to occupants from formaldehyde.

> The cleansing ability of wool for indoor air has successfully been demonstrated through numerous field experiments over a period of 7 years. Resin-based wood products, chipboard in particular, are the most significant sources of formaldehyde in indoor air. The urea-formaldehyde-resins (UFR) make up most of the resins used. UFR are subject to a slowly progressing hydrolysis, in the course of which formaldehyde is released as a function of air humidity. According to current estimates the German threshold value for formaldehyde in indoor air (0.1 ppm) is exceeded in approx. 10% of German households. This means that potentially several million people are affected. The cleansing ability of wool for indoor air has successfully been demonstrated through numerous

> field experiments. Results for air quality restoration in official buildings, such as kindergartens, by incorporation of wool continue to show drastically and persistently lowered formaldehyde concentrations.[18]

Wool is characterised as a reactive absorbent. Its rate of absorption is also affected by temperature. Absorption of formaldehyde is taken up at a faster rate the higher the humidity in the building, so moisture vapour acts as a carrier for the formaldehyde. The proteins in the wool consist of amino acids. Aldehyde emissions react with the amino acids, leading to what is known as a *stable methylene bridge*. This has the effect of binding the aldehydes into the material and converting them into a stable and less toxic chemical.[19]

Other scientists have recorded similar findings that wool fibre can reduce indoor formaldehyde.

> The data provided also show that wool fibre is capable of sorbing significant amounts of vapour phase formaldehyde from air, although some may be re-released under certain conditions. These data imply that wool used in textiles and insulation may be a realistic method of reducing airborne formaldehyde contamination.[20]

Wool has been tested as a suitable filter material for air cleaners in air handling systems, etc.

> In active filtration systems, wool can effectively reduce formaldehyde concentrations between 60–80%, depending on filtration fluid mechanics and room parameters. However, the effective capacity of the wool varies significantly depending on how airflow patterns affect how formaldehyde in the air contacts reaction sites on wool.[21]

While sheep's wool can be an effective way of dealing with formaldehyde this facility may differ between different products. Some sheep wool insulation products contain significant

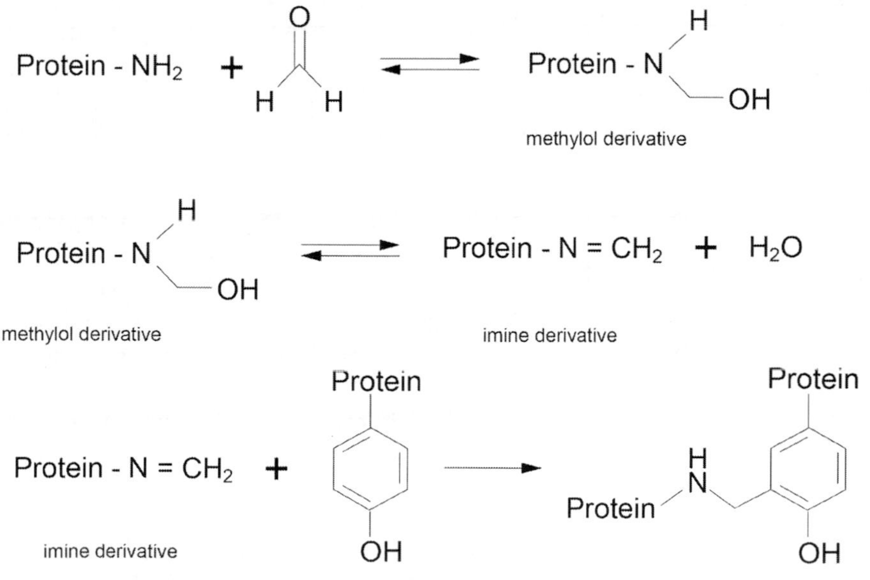

Protein - NH_2 + $H_2C{=}O$ ⇌ Protein - N(H)–CH_2–OH

methylol derivative

Protein - N(H)–CH_2–OH ⇌ Protein - N = CH_2 + H_2O

methylol derivative — imine derivative

Protein - N = CH_2 + Protein–C_6H_4–OH → Protein–NH–CH_2–C_6H_3(Protein)–OH

imine derivative

Figure 9.2 The protein process in wool absorption

amounts of synthetic glues and binders and polyester fibre and can be heavily dosed with chemicals to resist insects and fire. Sheep wool insulation products are now made in China where they claim 97 per cent formaldehyde and VOC absorption but they may contain a proportion of polyester.[22]

Only one sheep wool insulation product is certified by Natureplus as it contains minimal synthetic and toxic additives. This product is largely natural wool and can thus adsorb formaldehyde, whereas other products may be less effective. However if there is a barrier such as plasterboard/drywall between the insulation and the indoor air, this may be sufficient to reduce the adsorption of formaldehyde.[23]

> Wool removes harmful substances from the air: The secret behind this excellent ability is keratin, the protein building block which makes up sheep wool. The molecules in the side branches of the amino acids are able to absorb and neutralise harmful substances such as formaldehyde. This is done through a chemical reaction in which the harmful substances are connected to these molecules and converted.[24]

Sheep's wool has also been evaluated as a material to use for dealing with industrial wastewater![25]

German researchers Gabriele and Franz-Josef Wortmann discovered that, by applying wool to formaldehyde-contaminated houses and public buildings, the concentrations were reduced to less than half of the acceptable standards.

They showed that, when applying 300 ppm of formaldehyde to wool, about 97 per cent of the gas was absorbed by the fibre within 24 hours, with 80–88 per cent of the reduction having taken place in the first two hours. This research shows a significant reduction in formaldehyde emissions over a two-year period. There is a significant reduction initially and then a slower reduction over a long period of time. There is a difference between *physiorption*, which is fast, and *chemisorption*, which is relatively slow.[26]

A considerable amount of science has been devoted to how formaldehyde is modified by its cross linking with other substances. While some of this research seems inconclusive, it appears to be fairly certain that the formaldehyde is modified by the proteins in wool to a much less-hazardous chemical.[27]

Vapour permeability

It is important to ensure that fully breathable (vapour permeable) materials are used in a building, so that moisture, transporting toxic chemicals, can pass through layers such as plaster board (drywall). Unfortunately architects are increasingly specifying less breathable materials and building control officers are requiring non-permeable plastic layers to be used in the mistaken belief that this will reduce moisture problems. In reality using less permeable materials increases moisture problems. Vapour permeability can also be restricted by the use of plastic paint finishes.

Recent research at Bath University has suggested that bio-based, natural insulation materials have low toxic emissions and can absorb toxic chemicals. The Bath study looked at seven different bio-based insulation materials including wool.

> The use of natural bio-based insulation materials has increased in recent years, largely driven by concerns over the embodied energy and whole-life environmental impact of insulation materials. This has led to their increased use, typically within breathable wall

> constructions. A breathable wall construction allows the insulation material to directly contribute to the indoor air quality.
>
> It is the potential of these materials to adsorb or react with VOCs and formaldehyde, effectively removing them from the indoor environment that promises the greatest benefit of the bio-based insulation materials, regardless of their initial emissions.[28]

In addition to insulation, wool carpeting can have significant benefits in removing toxic chemicals, as has been confirmed in studies in New Zealand.

> Research by AgResearch scientists has revealed that wool carpets significantly improve indoor air quality by rapidly absorbing the common pollutants formaldehyde, sulphur dioxide and nitrogen oxides. Not only does wool neutralise these contaminants more quickly and completely than synthetic carpet fibres, wool does not re-emit them, even when heated. Wool carpet may continue purifying the air for up to 30 years.[29]

Sheep wool is confirmed, along with other bio-based insulation materials, as emitting very low levels of both toluene and formaldehyde, according to the Bath University research. This study looked at cellulose flakes, wool, hemp fibre, wood fibre, wood fibreboard and hemp lime/hempcrete. All had very low formaldehyde emissions apart from the cellulose flakes and wood fibreboard, which were a little higher.

> Under standardized room conditions, the emissions are not expected to cause any impacts on health and wellbeing based on current guidelines. A single wood-fibre based insulation did show a significantly higher emission of VOCs that would potentially raise concerns under certain circumstances.[30]

Studies have also been carried out on wool carpet to remove contaminants. 'Combining an ability to actively buffer moisture changes in the indoor environment, with a considerable capacity for absorbing and retaining indoor air pollutants, wool carpets provide a natural means of improving and maintaining indoor air quality'.[31]

Wool products have been used in Germany to reduce formaldehyde in schools. 'Sheep in School: From new construction to renovation case: A few weeks after first time use of a new School building complained children on complaints. Cause was too high a concentration of formaldehyde. Sheep wool came to the rescue'.[32]

Scavenging aldehydes using other materials

A number of mainstream construction materials manufacturers have recently spotted a market opportunity in promoting healthy buildings. One example is St. Gobain, a French-based multinational company that is one of the world's largest manufacturers of glazing and glass products. St. Gobain is also one of the world's largest producers of fibreglass insulation products, best known through the name Isover. The company has grown rapidly since the 1990s, acquiring UK builders merchants like Jewsons and taking over British Gypsum, and is one of the largest manufacturers of plasterboard and similar products. Recently St. Gobain has been promoting what it calls the 'multi-comfort' concept. Multi-comfort is promoted as achieving PH standards but also taking a 'holistic approach to comfort health and wellbeing in buildings'. Multi-comfort documents refer to some of the issues discussed in this book and suggest that good indoor air conditions can be achieved through the use of

St. Gobain construction products and mechanical ventilation.

> Saint-Gobain offers several product categories that have a direct impact on indoor air quality: products with the lowest emissivity possible for the building envelope in insulation, dry lining, facade, wall or floor covering; products contributing to the performance of ventilation systems such as high performance windows, doors, and technical insulation; products that help purify indoor air by scavenging certain COVs [sic], namely Aldehydes.[33]

The claim by St. Gobain that the multi-comfort home uses materials that can 'scavenge' VOCs and aldehydes is an attractive concept but is so far unsubstantiated by any real evidence. Temprell states that the building materials they use can 'decompose internal pollutants into non-harmful inert compounds' but no details are provided on how the building products will achieve the scavenging performance.[34]

St. Gobain has issued glossy literature about the multi-comfort claims and has supplied materials free to a couple of demonstration projects. One of these, an independent school sports hall at Kings Hawford (fees of £4,000 per annum) in Worcestershire received up to £500,000 worth of free St. Gobain materials and was used as a venue to launch the multi-comfort concept in May 2106.

It was necessary to look at documents from the USA to get details of the St. Gobain products, which they claim can 'absorb' formaldehyde. One St. Gobain product marketed in the USA is called 'Novelio', which is a glass fibre wall covering that they claim can 'capture 70% of formaldehyde as soon as applied'. They claim it remains effective even after several layers of paint have been applied.[35] It appears that this product was introduced to the UK in 2015[36] but the St. Gobain website is only promoting it as a reinforcing material, and not one to absorb formaldehyde. St. Gobain refers to tests carried out by EUROFINs that confirm the 70 per cent absorption claim but this was in a document published by St. Gobain. St. Gobain US subsidiary Certainteed also launched a gypsum wall board called Air Renew on the market in the USA in 2011, which they say absorbs formaldehyde but this was treated with some skepticism by Melton and Wilson of Building Green who were unable to find out the nature of the active ingredient in the pending patent.

> While we at EBN are excited about the prospect of a relatively standard building material removing a significant indoor contaminant, the proof is in the pudding. We look forward to reviewing test data that demonstrates reduced formaldehyde levels in indoor air during actual use over a long period of time.[37]
>
> In addition to CertainTeed's assurances that formaldehyde is not re-remitted once absorbed, AirRenew does carry GreenGuard certifications and a UL validation that the boards are capable of 'permanent formaldehyde absorption capacity of 0.4g/m^2 of surface area.'[6] But, these third party evaluations don't necessarily put concerns about scavengers potentially re-emitting formaldehyde in the future to rest … They could also end up being just another form of greenwash. While AirRenew certainly seems to be a promising application of formaldehyde scavengers, greater transparency and more research into on-site performance is needed, leaving us with lots of questions about these products. We encourage CertainTeed to disclose more about their additives, and about the compound formaldehyde is converted into before being absorbed by the boards.[38]

In a St. Gobain AirRenew brochure section entitled 'How Does it Work', it simply says that the gypsum board's 'light blue core' captures and transforms formaldehyde without providing any information about how it works![39] These claims are validated by a company called UL Environment Inc.[40] This is an interesting organisation that provides testing and environmental certification to industry in the USA and according to third party websites, UL now owns Greenguard which is one of the leading environmental certification bodies in the USA, closely linked to the USGBC and LEED.

Ironically, while attempts are being made to introduce formaldehyde-scavenging chemicals into gypsum board products, gypsum boards themselves may have been made with a potentially hazardous additives, including formaldehyde. One of the additives that has been used for some years is Ca-polynaphthalene sulphonate, which is used as a liquefier/dispersant. Ca-polynaphthalene sulfonate contains formaldehyde, which is released during the manufacture of the wall boards and napthalene is also a hazardous chemical. Apparently attempts have been made to substitute similar products that release less formaldehyde. As manufacturers are evasive about releasing details of their products, it has been difficult to ascertain which gypsum boards in the UK contain any of these chemicals.[41]

Certainteed in the USA sells insulation board products that contain phenol formaldehyde according to the Green Home guide.[42]

Scavengers for wood products

Formaldehyde scavengers have been applied to wood products for some years using a range of chemical additives such as ammonium phosphate and sodium metabisulphite. Ammonium phosphate should also be seen as a hazardous material, though possibly not as dangerous as formaldehyde.[43]

Research on the use of natural tannins (which occur in wood anyway) as a scavenger for formaldehyde has also been carried out, but it is not clear whether this is used commercially to any extent.[44]

Synthetic resins and glues are widely used in composite building products and some work has investigated the potential for natural materials to reduce formaldehyde.

> Four varieties of bio-scavengers, tannin powder, wheat flour, rice husk flour, and charcoal, were added to MF resin at 5 wt.-%. To determine formaldehyde emission and bonding strength, we manufactured engineered floorings. MF-charcoal was most effective in reducing formaldehyde emission because of its porous nature, but its bonding strength was decreased. Tannin powder and wheat flour, which contain more hydroxyl groups, showed higher bonding strength and curing degree than pure MF resin did.[45]

Some of the major companies operating in the global formaldehyde scavengers market are StarChem LLC, Georgia-Pacific Chemicals LLC, Emerald Performance Materials LLC and Dow Chemical Company among others.[46]

Fabric layers to remove indoor air pollution

More benign and less chemically aggressive methods for dealing with emissions exist, such as a product that has been developed in Sweden called cTrap,[47] which is a multilayer cloth intended to eliminate emissions in buildings with VOC emissions and moisture-related problems such as mould. The cloth has four layers, an inner protective layer, an outer protective layer, an adsorption layer and a semipermeable barrier, allowing absorption from the inside only. The two protective sheets are made of nonwoven polyester fabric surrounding the adsorption layer and a hydrophilic polymer sheet. It is designed to be fixed to walls behind thin plasterboard, using an adhesive tape, or laid under laminate floors or in ceilings. Apparently the adsorption layer is made of a carbon compound similar to the material used in water filters. The claims for the product are supported by the Swedish Research Council for Environment, Agriculture Science and Spatial Planning (FORMAS) and Lund University. It could be a very useful solution to dealing with hazardous emissions in existing buildings.

> cTrap was used to improve the indoor air quality of a school building with elevated air concentrations of 2-ethyl-1-hexanol. A clear improvement of the perceived air quality was noticed a few days after a cTrap prototype had been attached on the PVC flooring. In parallel, decreased air concentrations of 2-ethyl-1-hexanol were found as well as a linear increase of the amounts of the same compound adsorbed on the installed cTrap cloth as observed up to 13 months after installation.[48]

Measuring emission rates from materials

Much of the scientific literature in this area is concerned with measuring emissions rates of materials.[49] Adsorption is used in laboratory experiments, not to deal with reducing the problem of VOCs, but to measure how they are emitted. This laboratory research is valuable because it confirms how easily dangerous levels of chemicals are released. Commercial companies that are responsible for materials that produce VOCs and other emissions tend to take the line that emission levels are so low that there is not really a serious health problem. However scientific literature on what are known as 'secondary emissions' seems to confirm overwhelmingly that emissions rates can be surprisingly high, even when a material has been in place in a building for many months or even years. Emission rates do fall, especially if ventilation rates are high, but this may not be the case in increasingly airtight buildings.

> Most inner surfaces may store chemicals and equilibrium with the surrounding air will be approached by desorption or adsorption of airborne chemicals … adsorbed pollutants may diffuse into materials and their reemission is typically slower and also influenced by the speed of diffusion back to the surface.[50]

Some literature on green building refers to the value of chemical sinks but in the past has given little guidance on how this takes place or what materials can be used. Spiegel and Meadows[51] point out that all surface materials will absorb chemical molecules but then suggest that these are then released again later. Indeed there is some confusion in the earlier literature between adsorption that is beneficial and the way in which materials absorb toxic chemicals that then remain the building as a problem. It is important not to confuse these two. Organic compounds originate from a variety of indoor sources and outdoors. Hundreds of organic chemicals occur in indoor air, although most are present at very low concentrations. These can be grouped according to their boiling points and one widely used classification identifies four groups of organic compounds: (1) very volatile organic compounds (VVOC); (2) volatile (VOC); (3) semi-volatile (SVOC); and (4) organic compounds associated with particulate matter (POM). Particle-phase organics are dissolved in or adsorbed on particulate matter. They may occur in both the vapor and particle phase depending on their volatility. For example, PAHs consisting of two fused benzene rings (e.g. naphthalene) are found principally in the vapor phase and those consisting of five rings (e.g. benz[a]pyrene) are found predominantly in the particle phase.[52]

Plants

It is not uncommon to find publications and websites promoting plants as a way of dealing with indoor air problems. It is claimed that plants can clean air and absorb pollutants. Interior landscaping companies make extravagant claims about the effectiveness of spider plants, philodendron and golden pothos to remove 80 per cent of formaldehyde in 24 hours based on research by NASA.[53] However research has shown that the soil bacteria in the plant pots does as much, if not more, as the plant leaves, to absorb pollutants. Activated carbon has also been added to soil and appears to absorb pollutants.[54] The NASA study, which did not actually reach a firm conclusion about the ability of house plants to remove pollutants, is frequently cited by commercial businesses promoting the use of house plants.

> Living, green plants can remove toxic chemicals including formaldehyde, benzene and carbon monoxide from the air, according to a two-year study by NASA scientists. The following plants are particularly effective: areca palm, lady palm, rubber plant, dragon plant, English ivy, peace lily, gerbera daisy, snake plant, spider plant, weeping fig.[55]

> A University of Technology, Sydney (UTS) study with 60 offices in three buildings, found that three Dracaena deremensis 'Janet Craig' (300 mm pots), or Spathiphyllum wallisii 'Sweet Chico' (200 mm pots): –reduced total volatile organic compound (TVOC) levels by up to 75% and –lowered CO_2 levels by from 10 to 25% and –carbon monoxide (CO) levels by 90%.[56]

> UTS laboratory research (17 species/ varieties) shows indoor plants: Have strong capacity to remove VOCs; and If concentrations rise, so do rates of removal. All species are about equally effective – main removal agents are normal root-zone bacteria; plant nourishes and regulates its microorganisms (symbiosis) VOCs are removed day and night (24/7). Absorbed VOCs don't accumulate – broken down to CO_2 and water 3 to 6 plants/ office kept VOC levels below 100 parts per billion (ppb) regarded as negligible health risk (Aust. recommended total VOC max. is 500 ppb).[57]

Much of the work claiming that house plants could help with IAQ was carried out in the 1990s, such as a study at BRE.[58] However, as with much of the information on similar websites, it was impossible to find any significant scientific sources to support the claims. There is a considerable amount of scientific literature on how plants clear up toxins in the ground however. Despite the lack of evidence, using plants to assist in toxic emission mitigation cannot do any harm.

Notes

1 Parliament restoration plan could cost up to £5.7bn, BBC News www.bbc.co.uk/news/uk-politics-33184160 [viewed 10.3.16]

2 Dirty care home smelled so bad inspectors picked up 'offensive odour' OUTSIDE the building, *Manchester Evening News* www.manchestereveningnews.co.uk/news/greater-manchester-news/dirty-care-home-smelled-bad-8430358 [viewed 10.3.16]
3 Ford, S. (2015) *Nursing Times* www.nursingtimes.net/clinical-subjects/respiratory/air-quality-in-nursing-homes-affects-lung-health-of-residents/5083255.fullarticle [viewed 10.3.16]
4 Bentayeb, M. *et al.* (2015) Indoor air quality, ventilation and respiratory health in elderly residents living in nursing homes in Europe. *European Respiratory Journal* http://erj.ersjournals.com/content/early/2015/03/11/09031936.00082414.full.pdf [viewed 10.3.16]
5 Bentayeb *et al.* (2015) op cit.
6 Norwood, J. (2015) Sickening smell inside Cary Elementary School concerns parents http://wtvr.com/2015/04/30/cary-elementary-school-odor/ [viewed 10.3.16]
7 Bayer, C.W., Crow, S.A. and Fischer, J. (2000) *Causes of Indoor Air Quality Problems in Schools Summary of Scientific Research*, US Department of Energy ORNL/M-6633/R1
8 Healthy Schools Network (2012) Parent's Guide to School Indoor Air Quality, Healthy Schools Network
9 Bradley, J. (2016) Indoor Air Quality Issues in Schools: How Indoor Pollution Affects Health and Learning www.linkedin.com/pulse/indoor-air-quality-issues-schools-how-pollution-affects-jeff-bradley [viewed 10.3.16]
10 Air Quality in Schools, Vermont Department of Health http://healthvermont.gov/enviro/indoor_air/air_school.aspx [viewed 10.3.16]
11 Jones, B.M and Kirby, R., Indoor Air Quality in UK school classrooms ventilated by natural ventilation windcatchers, The Bartlett School of Graduate Studies, University College London http://discovery.ucl.ac.uk/1346486/2/Benjamin%20Jones%20-%20Paper%20and%20Plots.pdf [viewed 10.3.16]
12 Report reveals few Local Authorities know if schools comply with IAQ standards, Camfil www.keepthecityout.co.uk/2012/11/report-reveals-few-local-authorities-know-if-schools-comply-with-indoor-air-quality-standards/#sthash.Bqvzhp9c.dpuf [viewed 10.3.16]
13 Ajiboye, P., White, M., Graves, H. and Ross, D. (2006) *Ventilation and Indoor Air Quality in Schools*, Building Research Establishment for Office of the Deputy Prime Minister: London, Guidance Report 202825, Building Research Technical Report 20/2005 March 2006
14 Iddon, C.R. and Hudleston, N. (2014) Poor indoor air quality measured in UK classrooms, increasing the risk of reduced pupil academic performance and health. Proceedings of the international indoor air conference 2014, Hong Kong
15 www.eveningexpress.co.uk/fp/news/uk/did-40-children-fall-ill-at-school-because-of-mass-hysteria/ [viewed 10.3.16]
16 Däumling, C. (2011) Umweltbundesamt UBA Certech, Information from Powerpoint Presentation
17 Däumling, C. (2013) Hazardous Chemicals in Products, The need for enhanced EU regulations, Brussels, 29 October www.anec.eu/attachments/UBA_Indoor%20emissions_ANEC-ASI%20CC%20conference%202013.pdf [viewed 10.3.16]
18 Wortmann, G., Thome, S., Fohles, J. and Wortmann, F.J. (2005) Sorption of aledehydes from indoor air by wool, Formaldehyde as an example, 11th International Wool Research Conference, Leeds, 4–9 September
19 Wortmann *et al.* (2005) op cit.
20 Curling, S.F., Loxton C. and Ormondroyd, G.A. (2012) A rapid method for investigating the absorption of formaldehyde from air by wool, *Journal of Material Science* 47, 3248–3251
21 Wang, J. (2014) Assessing Sheep's Wool as a Filtration Material for the Removal of Formaldehyde in the Indoor Environment, MSc, University of Texas at Austin
22 KPS, sheep wool thermal acoustic mats promotional leaflet
23 Natureplus Awards Guideline GL0103, Sheep's Wool Insulation September 2010
24 Isolenawolle Sheep Wool Insulation brochure www.isolena.at/file/385/download?token=TSR02xBs [viewed 10.3.16]

25 Ghanbarnejad, P., Goli, A., Bayat, B., Barzkar, H., Talaiekhozani, A., Bagheri, M. and Alaee, S. (2014) Evaluation of Formaldehyde Adsorption by Human Hair and Sheep Wool in Industrial Wastewater with High Concentration, *Journal of Environmental Treatment Techniques* 2 (1), 12–17

26 Wortmann. *et al.* (2005) op cit.

27 Metz, B. *et al.* (2004) Identification of Formaldehyde-induced Modifications in Proteins: Reactions with model peptides, *Journal of Biological Chemistry* 279 (8) Issue of February 20, 6235–6243, 10.1074/jbc.M310752200

28 Maskell, D. *et al.* (2015) Properties of bio-based insulation materials and their potential impact on indoor air quality, First International Conference on Bio-based Building Materials, June 22nd–24th, Clermont-Ferrand, France

29 McNeil, S. (2014) Removal of Indoor Air Contaminants by Wool Carpet, *Technical Bulletin*, AgResearch

30 Maskell, *et al.* (2015) op cit.

31 McNeil (2014) op cit.

32 Schrader, S., Formaldehydsanierung Schafe in der Schule www.fisolan.ch/frontend/documents/dynamic/Schulsanierung_Formaldehyd.pdf [viewed 10.3.16]

33 www.multicomfort.co.uk/sectors/indoor-air/ [viewed 10.3.16]

34 Temprell, S. (2015) Home Comforts, *ShowHouse* p. 53

35 Clean Air www.adfors.com/us/brands/novelio/cleanair [viewed 10.3.16]

36 Adfors brings wall protection Novelio to the UK market www.sleepermagazine.com/adfors-brings-wall-protection-novelio-to-the-uk-market/ [viewed 10.3.16]

37 Melton, P., Wilson, A., Air Renew Wallboard Absorbs Formaldehyde from Indoor Air www2.buildinggreen.com/article/air-renew-wallboard-absorbs-formaldehyde-indoor-air (viewed 10.03.16)

38 Coffin, M. (2013) Hungry Building Products? www.pharosproject.net/blog/show/174/hungry-building-products [viewed 10.3.16]

39 UL Environment Validates CertainTeed AirRenew, Indoor Air Quality Gypsum Board Product Family for Environmental Claims 08/28/2012 www.certainteed.com/pressroom/pressRelease.aspx?id=601 [viewed 10.3.16]

40 Environment UL helps manufacturers capture value for their sustainability efforts http://industries.ul.com/environment [viewed 10.3.16]

41 Interactions between admixtures in wallboard production, Global Gypsum www.globalgypsum.com/magazine/articles/670-interactions-between-admixtures-in-wallboard-production [viewed 10.3.16]

42 http://greenhomeguide.com/askapro/question/does-insulation-containing-formaldehyde-continue-to-offgas-into-the-home-after-the-walls-are-sheetrocked-and-painted [viewed 10.3.16]

43 www.google.com/patents/US20110171482 [viewed 10.3.16]

44 www.wki.fraunhofer.de/en/services/vst/projects/reducing-HCHO-wood-based_materials.html [viewed 10.3.16]

45 Kim, S., Kim, H.-J., Kim, H.-S. and Lee, H. H. (2006) Effect of Bio-Scavengers on the Curing Behavior and Bonding Properties of Melamine-Formaldehyde Resins, *Macromolecular Materials and Engineering* 291 (9), 1027–1034

46 Formaldehyde Scavengers Market: Global Industry Analysis, Size, Share, Trends and Forecast 2014–2020 www.transparencymarketresearch.com/formaldehyde-scavengers-market.html [viewed 10.3.16]

47 www.ctrap.se [viewed 10.3.16]

48 Larsson, L., Markowicz, P. and Mattsson, J. (2015) Improving the indoor air quality by efficient exposure reduction: the surface emissions trap. Lund University, Sweden, Healthy Buildings Europe Conference, 18–20 May, Eindhoven

49 Lee, C.S., Haghighat, F. and Ghaly, W.S. (2005) A study on VOC source and sink behavior in porous building materials: analytical model development and assessment. *Indoor Air* 15 (3), 183–196

50 Salthammer, T. (2009) *Organic Indoor Air Pollutants: Occurence, Measurement, Evaluation*, 2nd Edition, Wiley

51 Spiegel, R. and Meadows, D. (2010) *Green Building Materials: A Guide to Product Selection and Specification*, 3rd Edition, Wiley

52 Crump, D. (2011) Nature and Sources of Indoor Chemical Contaminants. In Guardino Solá, Xavier (ed.), *Encyclopedia of Occupational Health and Safety*, International Labor Organization, Geneva

53 http://sunsethillsfoliage.com/clean_air.html [viewed 10.3.16]

54 Wolverton, B.C. *et al.* (1989) *Interior plants for indoor air pollution abatement final report*, National Aeronautics and Space Administration Stennis Space Centre Mississippi

55 Sevier, L. (2009) How to Improve Indoor Air Quality, *The Ecologist*, March 3 www.theecologist.org/green_green_living/how_to/269613/how_to_improve_indoor_air_quality.html [viewed 10.3.16]

56 http://greenplantsforgreenbuildings.org/news/potted-plants-can-significantly-reduce-urbanindoor-air-pollution/ [viewed 10.3.16]

57 www.uts.edu.au/sites/.../indoor_plant_brochure_2014.pdf [viewed 10.3.16]

58 Coward, M., Ross, D. and Coward, S. *et al.* (1996) Pilot Study to Assess the Impact of Green. Plants on NO_2 Levels in Homes, Building Research Establishment Note N154/96, Watford

10 Healthy building theories

There are many different interpretations of the idea of healthy buildings. At an international level there has been a huge amount of interest in the subject, IAQ and related topics. This has provided an almost inexhaustible source of information for this book. However the level of knowledge and understanding of healthy buildings is barely on the agenda in the UK and Ireland, and when it is discussed, the issue of building materials and hazardous emissions is rarely considered. In order to understand why this is the case, it is necessary look at the different policies and theories adopted by a range of organisations and individuals and explore why so many have their head in the sand about the risks and dangers to building occupants.

Following the decline in interest in SBS, green, sustainable architecture and construction have been largely fixated on energy efficiency and reducing CO_2, often incorrectly referred to as *carbon* emissions. There is little doubt that there is a significant section of the professional community interested in sustainable construction, but energy efficiency is all that is considered at the expense of all other issues. Adopting a more holistic approach to consider embodied energy, environmental impact such as pollution and health is frequently put to one side. This is justified on the basis that building owners and occupants are only interested in saving money by reducing their energy bills, or that reducing energy consumption will be the only way to 'save the planet'.

Despite this rather narrow energy-fixated view of sustainable buildings, there have been plenty of leading figures to adopt a more holistic approach in the past. Some early pioneers of green building were convinced of the importance of health, even if they were not always able to put it into practice through the technology available at the time. Ton Alberts, architect of the NMB bank building in Amsterdam was quoted in an important book, *Healthy Buildings*, in 1992: 'a building has to be as comfortable as our skin: it lives, it breathes, it surrounds us without oppressing us. A building has to be as healthy as we want our body to be'.[1] It was claimed that absenteeism among workers at the NMB building was 15 per cent lower than in their old building, though we have no idea how high it was there. This is due to it being a 'healthy and attractive building'.[2]

Architects and green building practitioners had access to information about buildings and health as early as 1990 through the Rosehaugh guide, *Buildings and Health*. This was a massive and comprehensive analytical guide of 522 pages that said that, 'individuals and organisations can be persuaded that their actions can ... provide a less hazardous more comfortable built environment in a renewable or sustainable manner'.[3] This guide was cautious, carefully researched and adopted a non-radical approach, and was funded by property developer, Rosehaugh (though Rosehaugh went bust in 1992!).[4] Despite its mainstream respectability, the book was largely ignored by the profession and industry for 25 years. Concerns raised about the health risks of manmade mineral fibre insulation materials and a range of toxic emissions had little impact on the greater uptake of hazardous materials by architects and specifiers. One possible explanation for this was the perception, at the time, that both holistic and green building design approaches were too heavily associated with alternative and 'hippie' concepts of architecture and building. 'Nostalgia for the Hippie Building Heyday: Today's green building movement traces back its roots (in part) to the improvised shelters built by back-to-the-land hippies in the 1970s'.[5]

The alternative green architecture movement was heavily influenced by New Age and mystical beliefs, which led to the hippie, lentils and sandals image. Some of the early pioneers of green architecture associated themselves with alternative culture and lifestyles. A critique of the failure of much early literature to make a distinction between 'objective reality and mystical beliefs' offended many green and eco architects who firmly believed that mystical ideas such as Feng Shui, geopathic stress and ley lines were scientific.[6]

Pearson, who set up the Ecological Design Association, advocated the idea of 'healing architecture'[7] in his *Natural House Book*, and promoted the idea of eco-architecture as being a path to *spiritual awareness* in a later book, and for some this is still the basis of the idea of healthy buildings.[8] In another similar book, *The Healthy House*, the authors, Baggs and Baggs, tried to persuade us that Feng Shui is synonymous with more scientific holistic design principles. Apart from referring to design rules formulated by Yan Yun Sung from AD 840–888, they also draw on astrological 'principles', linking these with building biology and some perfectly sensible scientific principles, such as using hygroscopic materials to control moisture.[9]

While some eco-architects have genuinely held New Age religious and mystical beliefs, the association of these with more scientific notions of healthy buildings did little to help the cause of widening interest in mainstream construction.[10] This may have scared off many architects and builders, due to a confusion between alternative medicine and healthy building ideas. It has taken some time for a rational scientific approach to healthy buildings to recover from this. Many feel great benefits from alternative and natural health techniques and lifestyles, some of which can have a sound scientific basis, but there is much to done to distinguish pseudo-science from sound alternative medicine. The debate will continue for many years, as has been the case in places like Totnes, Dartington and the Schumacher College, for instance.[11]

It is important that healthy building theory is based on solid science rather than mystical beliefs. However, those who seek an alternative to a technology-based society should find much of what is in this book useful to support their case. Good science must be *evidence based* and this requires research and testing, much of which is still absent from the healthy building world. Singh and Ernst explored this issue in terms of alternative medicine and recognise that some alternative treatments, once regarded as maverick, become mainstream once evidence proves they work effectively.[12]

Alternatives discussed in this book, such as natural and non-toxic materials, should not be seen in the same terms as homeopathy and crystal healing, but as practical solutions that have been used effectively. Many people are attracted by natural, healthier alternatives and some attribute this to the theory of 'biophilia', which is based on the idea that humans possess an innate tendency to seek connections with nature and other forms of life.[13] Biophilia should not be confused with biomimicry, which is a basis of inspiration for designers who try to mimic natural forms. This is also sometimes mixed up with healthy building science, but there may be interesting paths to explore by learning from nature, through biophilia and biomimicry.[14]

At the opposite end of the spectrum from naturopathy: there are also those who argue that technology and synthetic products are better for us than natural ones. This is justified by an extreme argument that natural organisms such as *botulinum* and *tetanospasmin* (that causes tetanus) are more dangerous than synthetic compounds: 'the most toxic natural chemical, *botulinum* toxin is over a million times more toxic than all of the synthetic chemicals, *except dioxin*'.[15]

Building biology

An institute of Building Biology was founded in Germany in 1983. This laid down many of the principles of healthy building, which have become widely accepted, and have influenced many of the practitioners of healthy buildings. However, even here, many of the practitioners of building biology (BB) also subscribe to New Age pseudo science. For instance the Australian College of Environmental Studies provides courses in 'Building Biology and Feng Shui' and a US BB practitioner calls their business, Feng Shui Magic.[16,17] Despite this, BB is a largely sound science and holistic approach to creating healthy and sustainable buildings.

> the main focus of Building Biology is human health…and achieving deep ecology is a bi-product of this. A central concept of Building Biology is that 'there is almost always a direct correlation between the biological compatibility of a building and its ecological performance.' In other words buildings that deeply nurture every aspect of human health in production, occupation, and post-habitation will also excel as models of sustainability. Why? … Because the natural environment is the gold standard for human health and the ultimate model of sustainability. The role of our indoor environments is to temper nature's extremes of temperature and to keep us dry and safe from predators.[18]

According to Healthy Building Science there are 25 principles of building biology, as shown in Figure 10.1.[19]

These principles are generally sound, though 'assuring low total moisture content' has led to a rigid rejection, by some German architects, of deep ecological building solutions such as hempcrete. Minimising the 'alteration of vital cosmic and terrestrial radiation' can lead to mystical beliefs in ley lines and geopathic stress.

Gale and Snowden Architects in Devon, England, are advocates of BB, though they also describe themselves as PH architects, which some might see as in contradiction with BB principles. They have established a UK Building Biology Association, which offers an online course that has been translated from the German Institut fur Baubiologie and Nachhaltigkeit. Some very useful material included is free to download, though not all of it has been translated into English.[20] They set out the following 'key elements' to be considered in BB, as well as the 25 principles:

- air – high IAQ
- water – clean water supplies and treatment
- materials – the use of non toxic and natural materials that are vapour permeable
- energy – low energy passive design and thermal comfort and the use of appropriate renewable technologies

BB investigates the indoor living environment for a variety of irritants including:

- VOCs in the air
- infestation from mould

- fungi and parasites
- daylight and acoustic levels
- electromagnetic radiation including: alternating electric fields, alternating magnetic fields, static electric fields, static magnetic fields, radio frequencies and radiation
- healthy occupation and maintenance of buildings[21]

1) Verify that the building site is geologically undisturbed
2) Place dwellings away from sources of air pollution and noise
3) Place dwellings well apart from each other in spaciously planned developments amidst green areas
4) Plan homes and developments taking into consideration the needs of the community, families, individuals and the natural ecosystem
5) Building activities shall promote health and social well-being
6) Use natural and unadulterated building materials.
7) Allow natural self-regulation of indoor air humidity using hygroscopic (moisture buffering) building materials
8) Assure low total moisture content and rapid desiccation of wet construction processes in new buildings
9) Design for a balance between heat storage and thermal insulation
10) Plan for optimal surface and air temperature
11) Provide for adequate natural ventilation
12) Use thermal radiation for heating buildings employing passive solar energy as much as possible
13) Provide ample natural light and use illumination and color in accordance with nature
14) Minimize the alteration of vital cosmic and terrestrial radiation
15) Minimize man-made electromagnetic and radio frequency exposure
16) Avoid building materials that have elevated radioactivity levels
17) Provide adequate protection from noise and infrasonic vibration or sound conducted through solids
18) Utilize building materials which have neutral or pleasant natural scents and which do not outgas toxics
19) Minimize occurrence of fungi, bacteria, dust and allergens
20) Provide the best possible water quality
21) Support building activities and production of materials which do not have adverse side effects and which promotes health and social well-being throughout their life-cycle
22) Minimize energy consumption utilizing renewable energy as much as possible
23) Source building materials locally and that do not contribute to the exploitation of scarce or hazardous resources
24) Utilize physiological and ergonomic knowledge in furniture and space design
25) Consider proportion, harmonic orders and shapes in design

Figure 10.1 The principles of building biology

Science-based policy on healthy buildings

There are a number of organisations in the USA that provide a comprehensive resource concerned with healthy buildings. The Healthy Building Network (HBN), founded in 2000, is perhaps the most important. Its aim is to reduce the use of hazardous chemicals in building products. The HBN points out that:

> Most chemicals used in building products are not tested for their impact on human health. We may assume everyday building products won't harm us, but we still can't reliably know that they won't. We focus on building products for two reasons:
>
> – First, the US EPA estimates that Americans spend as much as 90% of their time indoors, and can be exposed to unhealthy chemicals there at much greater concentrations than outdoors.
> – Second, the volumes of building products are so great – in excess of 3 billion pounds produced annually – that reducing the use of hazardous materials not only benefits the people who build, maintain and occupy buildings, but also the manufacturing sites and communities from which the materials come, and the dumps, incinerators or recycling facilities to which they are sent after their useful life.
>
> There are numerous examples – removing lead from paint, arsenic from pressure treated wood, and formaldehyde from particleboard – of how reducing the use of hazardous chemicals in our building products can improve the quality of our lives and our environment from the local to the global scale.[22]

Associated with HBN is the Pharos Project, which has created an independent database for identifying health hazards associated with building products.

> The Pharos Project encourages manufacturers to disclose all ingredients in building products; helps architects, designers and building owners avoid using products that contain harmful chemicals; and creates incentives for product redesign and modification to reduce the impacts of hazardous materials use throughout the lifecycle of building products.[23]

HBN/Pharos have issued a number of reports, which are free to download. They also have a subscription-based product library though there is a free trial period. They have a staff of a dozen or more and are well supported by architecture practices and a wide range of organisations. They appear to have a good working relationship with the US Green Building Council and some aspects of LEED. While they claim that their approach is entirely research based, it was not possible to find out the degree of independence they have from the companies whose products they include in their library.[24] HBN sets out the following principles:

> *The Right To Know.*
> We all have a right to know what is in the products we specify, buy and use.
>
> *Precautionary Principle.*
> Take precautionary actions based upon the weight of available evidence and in the face of uncertainty.

The Responsibility of the Manufacturer.
Manufacturers possess important information about the contents of their products and have a responsibility to be accountable for things they make.

Transparency.
Share all assumptions, methodology, data and analysis. Reward manufacturers who fully disclose contents and processes to allow for meaningful analysis.

Optimism.
Acknowledging that our goals are ambitious and difficult to attain, we believe they are within the grasp of committed professionals working in good faith.

Define the Ideal.
It is an act of optimism to set an ideal goal representing how we believe our products can be good for the world, rather than just issue prohibitions and warnings on what is bad.

Coalition and Consensus Building.
The sheer magnitude of tools, standards and ratings is now confusing and becoming counterproductive in the market place. HBN seeks consensus in establishing green materials standards.

Accessible Presentation.
Mindful of the complexity of the work we undertake, HBN will provide accurate materials that are elegant, informational and user-friendly.

Life Cycle Thinking.
Assess impacts along the entire life cycle of the material from extraction to disposal using a wide range of tools.[25]

Courses on healthy buildings

Derby University in England is the one UK university to offer a Masters course on 'Sustainable Architecture and Healthy Buildings'. It is recognised by the Chartered Institute of Architectural Technologists (CIAT).

> The programme has been designed as a contemporary sustainable architectural course with an innovative blend of established scientific theories and design principles in relation to the healthy buildings and environments, aiming to raise your awareness and practicing principles … You will also research issues that could adversely affect the health of the building users, in a people centered, proactive and preventive way.[26]

Other university programmes on sustainable architecture and building may touch on health issues but this is rarely mentioned explicitly. The Centre for Alternative Architecture Masters in Advanced Environmental and Energy Studies does include a module on occupant health and well-being.[27] Students are able to attend lectures in the Wales Institute for Sustainable Education (WISE), featured on the cover of this book.

In addition to the Institute of Building Biology, there are a number of organisations in Germany and Austria that have established clear science-based policies for healthy buildings and materials. The most important is Natureplus, The International Association for Future-Oriented Building and Accommodation, which is now established in several European countries.

> The European Association – Natureplus – promotes the use of building products, which have been strictly tested to ensure they do not cause any negative impacts on health. In order to achieve this aim, Natureplus created the quality label Natureplus®. Products which are awarded this label fulfill high standards relating to Climate protection, Healthy accommodation and Sustainability.[28]

The principles at the core of Natureplus certification of products are laid out.

> The special list of prohibited substances includes substances, according to CLP-Regulations, as per Directive 67/548/EEC, national law or classified by the named institutions as suspected of being carcinogenic, causing mutations or toxic to reproduction, toxic or sensitizing or classified as harmful to the environment. Furthermore, additional individual substances may be specified by Natureplus as non-desirable due to their environmental and health dangers and which one would not expect to find in a certified product.[29]

Another German organisation, Sentinel Haus, carries out testing of buildings, for which they provide certification and approve products using a 'traffic light' methodology: green, amber, brown and red. They rely heavily on Natureplus certification for materials and products.

> Sentinel Haus Institut helps to protect buildings from unwanted contaminants or structural damages. The word 'sentinel' derives from Middle English, meaning 'guard'. Just as a structural engineer guarantees that the house does not collapse, SHI stands for the innocuousness of indoor air. We train all protagonists of the construction process and bring you in contact with certified partners with respect to healthy housing. The Sentinel standard enables optimal IAQ with contractually stipulated security for your health. Sentinel Haus Institut supports you to create healthier buildings, whether that may be kindergartens, schools, hotels, retirement homes, work centres or your own private home, of course.[30]

An important Austrian initiative has led to the establishment of the Baubook, which is a database for ecological construction and renovation. So well accepted is this tool that local government subsidies are available in some regions of Austria if Baubook materials are used. The Baubook provides ready-made technical specifications that are used by local authorities and central government as part of green public procurement.[31,32]

Mainstream recognition of healthy building principles in Germany and Austria and to some extent, Switzerland, while not as extensive as some may like, has nevertheless given a huge economic boost to companies manufacturing healthy, non-hazardous building products. The lack of support for healthy building policies in principle in the UK, by contrast, means that such products have to be imported at a premium.

Notes

1 Holdsworth, B. and Sealey, A.F. (1992) *Healthy Buildings. A Design Primer for a Living Environment*. Longman
2 Edwards, B.W. (ed.) (1998) *Green Buildings Pay*. E&F Spon
3 Curwell, S., March, C. and Venables, R. (1990) *Buildings and Health*. RIBA Publications
4 www.independent.co.uk/news/business/city-hopes-dashed-as-rosehaugh-fails-1560855.html [viewed 8.3.16]
5 Holland, M. (2013) Musing of an Energy Nerd www.greenbuildingadvisor.com/blogs/dept/musings/nostalgia-hippie-building-heyday#ixzz42KKNDbZ3 [viewed 8.3.16]
6 Woolley, T. (1998) Green Architecture Man Myth or Magic?, *Environments By Design* 2 (2), 127–137
7 Pearson, D. (1989) *Natural House Book. Creating a Healthy, Harmonious and Ecologically Sound Home*. Gaia/Conran
8 Pearson, D. (1994) *Earth to Spirit In Search of Natural Architecture*. Gaia Books
9 Baggs, S. and Baggs, J. (1996) *Healthy House Creating a Safe Healthy and Environmentally Friendly Home*. Thames and Hudson
10 Huston, P. (1997) *Scams from the Great Beyond: How to make money off New Age Nonsense*. Paladin Press
11 Gray, J. (2012) The town that's twinned with Narnia – the capital city of pseudo science, *New Humanist*, September 4 https://newhumanist.org.uk/articles/2862/the-town-thats-twinned-with-narnia [viewed 7.3.16]
12 Singh, S. and Ernst, E. (2008) *Trick or Treatment: Alternative Medicine on Trial*. Bantam Press
13 Wilson, E.O (1986) *Biophilia*. Harvard College
14 http://healthybuildingscience.com/2016/02/17/the-science-of-biomimicry-how-to-use-natures-teachings-to-thrive/ [viewed 7.3.16]
15 http://blogs.scientificamerican.com/guest-blog/natural-vs-synthetic-chemicals-is-a-gray-matter/ [viewed 7.3.16]
16 www.aces.edu.au/courses/building-biology [viewed 7.3.16]
17 www.fengshuimagic.com/bau-biology/what-is-bau-biology/ [viewed 7.3.16]
18 www.motherearthnews.com/green-homes/building-biology-embracing-natures-wisdom.aspx [viewed 7.3.16]
19 http://healthybuildingscience.com/2013/02/05/building-biology-25-principles/ [viewed 7.3.16]
20 http://buildingbiology.co.uk/ [viewed 7.3.16]
21 http://ecodesign.s3.amazonaws.com/building_biology_dg_b60c4ff64f68f053.pdf [viewed 7.3.16]
22 www.healthybuilding.net/content/about-us [viewed 7.3.16]
23 www.healthybuilding.net/content/pharos-v3 ([viewed 7.3.16]
24 www.healthybuilding.net/content/research-and-reports [viewed 7.3.16]
25 www.healthybuilding.net/content/mission-and-principles [viewed 7.3.16])
26 www.derby.ac.uk/courses/postgraduate/sustainable-architecture-healthy-buildings-msc/ [viewed 7.3.16]
27 http://gse.cat.org.uk/msc-architecture-advanced-environmental-and-energy-studies [viewed 23.3.16]
28 www.natureplus.org/ [viewed 23.3.16]
29 Natureplus e.V. Award Guideline GL0000 BASIC CRITERIA Issued: May 2011 www.natureplus.org/fileadmin/user_upload/pdf/cert-criterias/RL00Basiskriterien_en.pdf [viewed 7.3.16]
30 www.sentinel-haus.eu/en/ [viewed 7.3.16]
31 http://alpstar-project.eu/uploads/cnaform/doc/5499/baubook%20info%20english.pdf [viewed 27.3.16]
32 www.innovationseeds.eu/Policy-Library/Core-Articles/The-Baubook-Portal.kl [viewed 27.3.16]

11 How to build healthier buildings

The main way to create healthier buildings and to reduce hazardous emissions and bad IAQ is to follow a number of simple design principles, which includes the precautionary principle. Even if you are not entirely convinced by the evidence in this book about the risks to health from hazardous, toxic and carcinogenic materials, it is still worth considering avoidance of risks. Building healthier buildings is not difficult, once architects and their clients accept that buildings do not need to be full of synthetic and hazardous chemicals. If natural and non-toxic materials are used instead, then hazardous emissions can be reduced significantly. The following basic ideas should not only be good for the health of building occupants but also good for the workers who produce building materials and the environment in general:

- *Structure*: Reduce masonry and concrete materials as much as possible where they are exposed to the indoor environment or can emit radiation into a building. Cement and concrete manufacture contributes as much as 10 per cent of CO_2 emissions and claims about the thermal mass benefits of concrete are misleading. Earth-based materials can be used instead. Use of recycled waste materials in concrete may not have significant environmental benefits and may contribute to hazardous emissions and radiation. Where concrete has to be used it should be well away from where it might affect indoor air. Ecological alternatives such as limecrete and rammed earth should be preferred.
- *Timber frames*: These should be used as much as possible. Radiation and cancer levels are much lower in countries where most people live in timber buildings. However care must be taken to ensure that chemical treatments for timber are not able to affect indoor air and considerable care must be taken to check the glues used in composite timber products. Timber frame and solid timber panels can be used for large multistorey buildings as well as houses.
- *Plasterboard and other lining materials*: Select dry walling materials that do not contain hazardous chemicals and additives and are as vapour permeable as possible. Ensure that timber panelling has not been treated with hazardous chemicals and made with hazardous glues.
- *Windows and doors*: Do not use uPVC windows and doors and restrict the use of powder coating where possible. Use solid natural wooden doors internally rather than composite acrylic-coated doors.
- *Insulation*: Use natural, renewable non-toxic insulation as much as possible. Where this is being used for its hygroscopic properties, ensure that the vapour permeability properties are not blocked by other materials. Do not use synthetic, petrochemical or mineral-based insulation materials where they can affect indoor air. It may be worth

considering excluding some synthetic materials from the outside of buildings due to their fire risk.

- *Airtightness*: Do not accept that wrapping buildings in plastic membranes with lots of adhesive tape is the best way to build. Even where membranes are claimed to be breathable, there can be problems with raised humidity and moisture. A good standard of airtightness can be achieved with careful detailing, wet plaster finishes and good quality timber windows and doors and the use of natural insulation materials such as those described below.
- *Ventilation*: While good natural (opening windows) and passive ventilation should be the preferred option, carefully selected mechanical ventilation may be necessary for wet rooms and some buildings. Dependence on ducted and heat recovery mechanical ventilation systems should be avoided at all costs as this will only become a maintenance headache and may add to indoor air problems. Moisture and dampness can be managed through natural and passive ventilation and the use of breathable and hygroscopic materials.
- *Paints and finishes*: Use natural paints, oils, stains and varnishes and check the constituents of so-called eco paints to ensure they don't include hazardous materials. Minimise the use of synthetic sealants and adhesives. Detail interiors to obviate their use.
- *Flooring*: Do not use PVC flooring finishes under any circumstances, selecting floor coverings made from rubber, linoleum, tiles, natural timber and bamboo and natural carpet, providing good environmental declarations are provided. Be careful with the use of laminated floorings; check glues that have been used. Ensure that glues and adhesives used with floor coverings do not include hazardous chemicals.
- *Furniture and indoor fittings*: Ensure that kitchen furniture and other internal fittings and fixtures made from composite wood products (as most are) have very low or zero formaldehyde and other hazardous emissions from glues and finishes. Avoid the use of uPVC/vinyl/acrylic-faced products. These are most commonplace today. Avoid the use of fibreglass, uPVC/vinyl/acrylic composite doors unless it is possible to obtain an assurance of zero emissions.

Timber frames and synthetic materials

While timber-frame construction should be the preferred option for healthy building, unfortunately the majority of timber-frame companies in the UK and Ireland are wedded to using synthetic insulation materials. Often their certification and warranties are also tied to the use of unhealthy and hazardous materials. First of all the structural integrity of stud timber frames is dependent on the use of composite timber boards such as OSB, which may contain hazardous glues. Also it is normal practice to wrap the whole frame in a membrane and to use synthetic insulation such as mineral fibre or plastic foams. The UK Structural Timber Association promotes the use of structural insulated panels (SIPs) using polyurethane foam (PU) or EPS.[1] There are other composite and 'sandwich' panel systems in use that mainly use mineral fibre. A significant number of timber-frame kit companies are owned by or closely associated with the leading manufacturers of synthetic insulation materials.

Timber frames and natural materials

It is possible to combine timber-frame construction with natural insulation materials and to cut out the use of plastic membranes and composite boards. Both post and beam and stud frame construction can be easily adapted to use a range of healthy insulation materials but it will be necessary to seek out timber-frame manufacturers that are willing to consider this.

Woodfibre

Perhaps the most widely used natural construction material with timber frame is wood fibre. This is an insulating board made from compressed wood material. The boards are made without or with minimum glues and additives and there are wide ranges of products available with different densities that can be used externally or internally and on roofs and floors.[2]

Some woodfibre products have Natureplus certification. It is possible to create an airtightness layer using wood fibre that avoids the need for plastic membranes, though many approved construction details continue to include the use of OSB racking boards. Wood fibreboards can also be used to renovate existing buildings and can meet most thermal performance requirements. Woodfibre-based boards are available that may be dense enough to provide racking resistance. All should be formaldehyde free. Check any products that may still use glues however.[3,4]

Hemp and wool insulation

There is a wide range of hemp fibre and sheep wool insulation products and these can be used instead of mineral fibre. Sadly few of these products are manufactured in the UK, and importing them from continental Europe into the UK and Ireland means increased costs. These products can easily match the thermal performance of synthetic insulation materials but have the added benefit of being vapour permeable and hygroscopic, so they can help to modify humidity and reduce the risk of dampness. Unfortunately it is not uncommon to see these materials used with conventional detailing, so that non-breathable materials enclose the insulation. It is important not simply to substitute these natural insulations but to detail the building to retain breathability.

There was some negative publicity about wool insulation becoming infested by moths a few years ago and some wool insulation products are treated with borax to resist this. Evidence of the alleged moth infestation has been hard to find and if it did occur it is likely to do with the material being used incorrectly and sealed into situations with high levels of moisture. There are examples of demonstration historic building renovation projects where wool has been used, trapped against existing damp walls with an inner lining of foil back PU insulation. This is not to be recommended. Suppliers of hemp and wool insulation products should be asked about the level of polyester or other synthetic glues and binders used. Wool insulation products are available with negligible hazardous content, but this does not apply to all products.

Case study of how wool insulation outperforms synthetic insulation

An interesting example of where a natural insulation product outperforms synthetic materials comes not from buildings but a branch of health and medicine: vaccine transportation. This provides striking evidence that natural materials do a better job than synthetics.

In order to deal with health problems in hotter developing countries, vaccines manufactured in western developed countries by large pharmaceutical companies have to be transported around the world. WHO has found that up to 50 per cent of vaccines are ineffective because of damage from getting too hot or too cold during transit or even in storage.[5]

It has been assumed that this problem occurred on the final stage of the delivery process as the vaccines may be carried to distant clinics on the back of mules or bicycles, whereas during the earlier stages of the journey, the vaccines are carried in climate-controlled trucks and containers. However the problem was found to be at the earliest stage in the transport process when they were chilled.

> Since most vaccines are sensitive to heat, an adequate cold-chain system often has to be created and maintained to preserve the quality of a vaccine before it is administered. Although emphasis has long been placed on avoiding high temperatures during vaccine storage and shipment, exposure to 'subzero' temperatures – i.e. temperatures <0°C – can also damage and reduce the potency of the diphtheria, tetanus and pertussis (DPT), diphtheria and tetanus, tetanus toxoid, hepatitis B and pentavalent vaccines.[6]

The degradation of vaccines may also happen when cartons are left sitting on pallets, waiting to be loaded into the climate-controlled containers. Thus the vaccines need to be wrapped in good insulation that will maintain a steady temperature whatever the external environment. To date, the bulk of vaccines are transported in polystyrene containers, as it was assumed that this provided good insulation. However it has been recognised that this is not the case and a UK company has come up with the idea of using sheep's wool insulation, as this not only works better as an insulator but also cushions the vaccines and is hygroscopic so it prevents moisture and condensation building up in the transported cartons. This is a clear and dramatic example of how natural insulation materials are better than synthetic.

> Woolcool insulated packaging has been proven to maintain stable internal temperatures between 2°C and 8°C in excess of 72 hours, more effectively than other packaging including polystyrene cartons, polyethylene foam or air pocket products. The felted wool insulation also provides more effective cushioning protection for fragile contents, such as vaccine vials.[7]

Hemp lime and hempcrete

Hemp mixed with a lime binder has become a well-established composite natural form of construction that provides a breathable form of insulation. Hempcrete has good thermal mass and is hygroscopic. It is normally mixed and cast on-site around or within a timber frame, using shuttering or by spraying. It is unpopular with some designers and contractors because it is a wet system, but it has the advantage of providing excellent airtightness and structural stability as it is not essential to use an OSB racking board and membranes can be entirely omitted. Hempcrete has been used in large multi-storey, commercial and industrial buildings as well as housing. It is possible to use hempcrete in prefabricated panels, though this lacks some of the benefits of casting or spraying on-site. While the lime binder can be regarded as caustic and thus must be handled with care, the final product has no hazardous emissions and the lime as a natural biocide can protect the timber frame from rot or infestation by insects.[8]

An example of the use of hempcrete in mainstream projects is by Marks and Spencer, the giant retail clothes and food group, which has adopted a green building policy and has used materials such as hempcrete in its new stores. An evaluation of the store at Cheshire Oaks, near Chester, is claimed to have a much lower energy consumption than a comparative benchmark store. 'The use of highly innovative building materials such as *hempcrete* and its exceptional air tightness ... has resulted in the store using 60% less heating fuel than predicted'.[9]

Cellulose insulation

Cellulose insulation made from recycled newspapers has had a mixed reception from those wanting to build healthy buildings. Highly dosed with borax salts for fire and pest resistance, this has limited its acceptance. However a cellulose insulation product is available from Austria that claims to be borate free and has a mineral fire retardant (8.4 per cent) that has allowed it to achieve Natureplus certification. However at the time of writing, information about the exact nature of the mineral fire retardant could not be obtained.[10] Despite concerns about the hazards of borax, cellulose insulation is widely used by ecological architects and builders, particularly in Germany and Austria. A study carried out in Ireland about the market potential for cellulose insulation concluded that it maintains a long-term healthy environment within a building and that sufficient underused waste-paper resources were available for it to be viable to manufacture in Ireland.[11]

Earthen materials

A range of materials and construction methods using unfired earth can be useful in the design of healthy buildings. Rammed earth and unfired (sometimes called adobe) earth bricks have been used successfully as thermal mass and earth can also help to buffer humidity. Cob construction of earth and a small amount of straw is also popular with self-build enthusiasts, but is largely a handcraft technique, unusual in mainstream construction. It is also possible to build with 'light clay'. There is a wide range of clay boards, earth plasters and clay-based paints and these can be very useful in finishing building interiors and retaining breathability. Advice and information can be obtained from Earth Building UK (EBUK).[12,13]

Strawbale construction

Building using strawbales is popular throughout the world and can be a practical solution for natural building in the UK. While it is largely associated with self-building, there are several buildings in the UK, both houses and public buildings, that include strawbale walls and have been built by contractors. Strawbale walls provide excellent thermal insulation, moisture buffering and breathability. Strawbale walls can be constructed as loadbearing (Nebraska style) or in various hybrid forms of construction where the strawbales are used in conjunction with timber frame, or made into prefabricated panels. Strawbale walls are normally finished with a lime render and it is important that breathable finishes are used.[14,15,16]

Plasterboard

As this book was going to press a new product appeared, soon to be available on the market, called Breathaboard. This is intended as an alternative to conventional gypsum boards and is made from natural fibre and cellulose materials. It is claimed to be both hygroscopic and able to absorb VOCs. Tests are currently being carried out at Bath University.

It is clear that many new 'bio-based' construction materials like Breathaboard are in development and that these are likely to provide robust and credible alternative to materials that contain hazardous chemicals. Insulation materials made from mushrooms are also in development, and a wide range of products based on bamboo, kenaf and other natural materials are already available. While mainstream construction will no doubt continue to be septical about such products, as research and standards are taken forward, they will eventually enjoy a greater share of the construction materials market.

Natural insulation materials can perform differently and better than synthetic materials

Not only are natural materials better for health, due to significantly lower levels of hazardous emissions, but they can also be more effective in improving and managing good IAQ, particularly in relation to moisture, humidity and dampness. The hygroscopic characteristics of natural materials were discussed in Chapter 7. This is not easy for many to understand, as conventional building physics does not deal adequately with these issues. Many advocates of energy efficiency insist on using petrochemical-based products.

> As I've stated more times than anyone here cares to remember, there is nothing either 'green' or healthy about petrochemical foams, or any of the 80,000 petrochemicals that never existed on earth before we created them. Unfortunately, most of (American) society is brainwashed into believing in the 'magic' of chemistry, as the advertisers and marketers have impressed on us for generations. Every product produced since the start of the petrochemical age is toxic, either to people or the environment or both. In spite of this growing understanding, most professionals in the 'green' building movement continue to rationalize the use of petrochemical materials in order to save petrochemical energy, further rationalized by the belief that it makes a house durable.[17]

It is often hard to get across to conventional professionals that natural materials have better thermal performance than synthetic because it involves a complete change of attitude. It is assumed that so-called 'high performance' synthetic insulations trap air and can out perform other more natural or traditional materials. While this may appear to be the case, when based on thermal resistance values established in laboratory hot box conditions, in real buildings the performance of synthetic lightweight materials is not always as good as predicted. Insulation performance claimed by synthetic product manufacturers is fed into energy prediction software such as the SAP to comply with building regulations but this may not reflect what happens in reality.[18]

Unintended consequences of airtight synthetic buildings

The use of synthetic and airtight materials in 'highly energy efficient' houses in particular has led to several unintended consequences. For instance there have been serious problems with buildings overheating.[19] Householders out at work all day come home to find their house has overheated during even a temperate summer in the UK. This is because the lightweight synthetic insulations not only fail to keep the heat in but they fail to keep the heat out.

Airtight buildings using synthetic and non-breathable materials can also trap moisture and this has led to serious problems of dampness, mould growth and poor IAQ. There is little doubt that the construction industry would dispute these allegations and there is continued support for the idea that the way to produce energy efficient homes in the future is using what has become known as modern methods of construction (MMC). However evidence is growing of recently built housing that has suffered from damp problems. A demonstration low-cost energy-efficient housing scheme in Milton Keynes, England (Oxley Woods, designed by Rogers Stirk Harbour and Partners and built by Taylor Wimpey), approved by NHBC, won over a dozen design awards, but is alleged to have developed serious problems with damp and defects.[20] [21] '"There is likely to be wet and dry rot and fungal infestation behind the rain screen of each dwelling, due to poor detailing and poor construction," the report states'.[22]

Another pioneering low-energy prefabricated scheme in Leeds known as CASPAR had to be demolished after it had been occupied for only five years.[23] The tabloid press has also drawn attention to problems with energy efficient homes. 'Are energy efficient homes making us ILL? Toxic mould caused by poor air circulation could trigger "sick building syndrome". It is thought to be present in 30–50 per cent of new or refurbished buildings'.[24]

While the *Daily Mail* may not be regarded as the most reliable source, scientific studies confirm these problems. A wide-ranging study by Shrubsole *et al.* refers to 100 unintended consequences of energy efficiency policies:

> links between lower air change rates and a rise in relative humidity (RH), leading to increases in house dust mites, mould, severity of asthma and allergies and in fabric decay in existing properties, particularly traditional buildings. In new builds, with tighter construction drying out times for 'wet trades' are extended leading to higher RH over a prolonged period during initial occupancy. Whilst a reduction in pollutants from external sources such as PM2.5, which has known negative health impacts is noted, an increase in exposure to indoor sourced pollutants such as PM2.5, volatile organic compounds (VOCs) and environmental tobacco smoke (ETS) may occur. There is also emerging evidence for a population-wide increase in cancer risk from increased exposure to radon indoors (an airborne pollutant known to be carcinogenic). Higher indoor temperatures can also lead to changes in indoor air quality through an increase in concentrations of indoor sourced pollutants, specifically VOCs.[25]

Opposition to the use of natural materials?

Despite problems with conventional petrochemical-based construction methods, there is a great deal of hostility to natural and ecological materials among the mainstream construction industry.

Natural materials are assumed to be:

- more expensive
- less durable
- difficult to build with
- lacking approvals and certification
- hard to obtain
- lacking credible environmental and health aspects

There is some truth in the assumption that many natural products are harder to obtain and more expensive. This is largely due to market forces. Sales and distribution of construction materials are largely in the hands of distributors, builders' merchants and DIY stores. They decide what the market needs and set the prices accordingly. Indeed, 98 per cent of all insulation sold in the UK is of petrochemical/synthetic materials and the government investigated allegations of price fixing and discounting of insulation materials a few years ago.

Many natural insulation materials such as sheep's wool, woodfibre and hemp are sourced from natural materials that are not in themselves expensive and if the market demand changed, they would be plentiful and could be sold much more cheaply. Waste wood is widely available and hemp could easily be grown as a worthwhile cash crop for farmers that fits easily into a crop rotation system.[26]

Bristow conducted a survey of architects and others in the Irish construction industry to assess their attitudes to the use of natural materials and provides confirmation of the negative attitudes that were based more on prejudice than real experience.

> For industry respondents, excessive cost was the most commonly named problem with natural building materials (NBMs) although in a separate quantitative question only 28% indicated that cost was a reason they had not used a natural building product. A second perceived problem that was highlighted by almost one third of respondents was that of the potential for decay and degradation, which would dissuade the specifier. Nevertheless, in a later question, only eight respondents admitted having or knowing of a negative experience with NBMs, and only seven described the problem. Almost one third of respondents did not believe that the environmental and health impact of material choice were as important to address as energy use and CO_2 emissions.[27]

Despite increasing mainstream adoption of natural materials such as woodfibre and hempcrete, the NHBC produced a highly damaging report about what they called cellulose-based materials in 2014. A foreword written by Rod MacEachrane, a retired director of the NHBC and a director of Barrat Developments from 2006,[28] did not reflect the content of the whole document, which was quite positive about natural materials.

> For such materials to become more widely accepted, it is important that they can be properly evaluated and assessed alongside other products and materials. It is important to emphasise that the current low incidence of use of the materials featured in this report in built properties means there is, as yet, relatively little statistically robust evidential data as to their long term performance.[29]

This was picked up by the construction press, leading to damaging headlines such as 'Builders warned of risks with natural materials'.[30] While the NHBC report was suggesting that more research was required to establish the longevity and robustness of natural materials, the misleading press reporting had a very damaging effect on the perception of natural materials in an industry that does not take time to read reports carefully. This may have contributed indirectly to the collapse of several companies producing natural materials and much harder market conditions for the producers of natural healthy materials in the UK.

Lime Technology, a company in Oxfordshire promoting hempcrete, has closed. Other examples of closures include the Black Mountain sheep wool insulation factory in North Wales and Excel Industries producing cellulose insulation in South Wales. Excel Industries, which used to produce Warmcel cellulose insulation in Wales, is now owned by CIUR, a company based in Prague, and they are offering a range of fibre-based products including cellulose.[31] Black Mountain sheep wool is still being advertised and is owned by International Petroleum Products. While this might seem a gloomy picture, it is still possible to source natural and ecological materials.

Availability and potential of natural construction materials

Sheep wool

Wool is controlled in the UK by the British Wool Marketing Board. Clip value prices for fleeces hover around £1 to £1.20 per kilo and it is well known that farmers are delighted if they can get more for their wool than it costs to shear the sheep. Global wool production is around 1.3 million tonnes per annum but less than 60 per cent of this goes into clothing manufacture. Some of the remainder goes into carpet production. CO_2 and methane emissions from sheep are frequently claimed to be high but this is exaggerated when compared with CO_2 emissions from petrochemical products. Sheep wool insulation products have been manufactured in the UK for some years by a number of companies and one in particular offers hemp insulation as well.[32] The only Natureplus-certified wool insulation, free of chemicals, is made in Austria. If demand for sheep wool insulation were to rise then there would be sufficient raw material to meet increased need, though it would take a while for manufacturing capacity to catch up. If sheep wool insulation were made in larger quantities it would undoubtedly fall in price.

Woodfibre products

Wood fibreboards and insulation batts are widely used and have been incorporated into mainstream construction products without difficulty in the UK. Unfortunately there is no manufacturing capacity in the UK and so products are imported from Switzerland, Germany and other European countries. Factories in England, Scotland and Ireland that process waste and low-value timber to make composite timber products such as OSB, MDF and other boards could easily convert or add production lines to make woodfibre insulation products. If this were to take place, the cost of woodfibre products would decrease significantly. They could be manufactured under license from other European manufacturers so as to ensure high standards and certification.

Hemp and lime

Hemp is grown and processed in vast quantities in mainland Europe. Hemp fibre is in great demand for bio-composite products and is used in the manufacture of car interiors. It is also grown and processed in the UK. One company in Yorkshire grows and processes its own hemp as the fibre is used in the manufacture of high-quality mattresses. The hemp 'shiv' that is required for building is thus readily available. Competition between a number of manufacturers in Europe of lime binders that can be used with hemp has resulted in a wide range of products being available, particularly in France where hempcrete is widely used and has government certification. A lime binder is also manufactured in Northern Ireland but most lime binders currently in use are imported from outside the UK. The raw materials for hempcrete are relatively economical and there are specialist subcontractors who work with hempcrete.

Strawbales

Straw is widely available but most strawbale construction uses small rectangular bales and these are less common on farms, which tend to make large round bales. Many people assume that strawbale construction is very cheap as the bales themselves are cheap when straw is readily available. However detailing a wall that is 500 mm thick either with or without a timber frame is a little more complex than many assume and this means that strawbale construction is only cheap if there is plenty of unpaid labour.

The potential for natural and renewable materials in construction

Currently the materials referred to above are still regarded with suspicion by many and marginal to mainstream construction at best. Changing supply chains to greater demand for natural materials will take time and thus increased demand for healthier buildings will create problems in terms of sourcing materials or increasing prices if they have to be imported from other European countries. However most natural materials are in themselves based on widely available resources such as waste wood and paper and crops that can easily be grown. The basic raw materials are cheap. What is lacking in the UK are policies and legislation that will force synthetic petrochemical materials off the market and encourage companies to invest in the processing and manufacturing of environmentally friendly, non-hazardous renewable materials. As Geiser says:

> The prospects for bio-based materials and processes, and an economy based more on renewable and recyclable materials, are dimmed by the current dominance of petrochemical materials. Without near term commercial prospects, market interest and private investment are likely to be slow to encourage these alternative materials and pathways. The detoxification and dematerialisation potentials of bio-processing will require more than current market incentives and current government policies.[33]

Notes

1 Structural Timber Association, SIPS www.structuraltimber.co.uk/timber-systems/sips [viewed 12.3.16]

2 Acara concepts, Pavatex www.acaraconcepts.com/wood-fibre-insulation/ [viewed 12.3.16]

3 Tradis, External Sheathing Board www.warmcel.co.uk/wp-content/uploads/2013/12/Excel-Timbervent-Technical-Data-Sheet-v8-12MM-v4.pdf [viewed 12.3.16]

4 Egger DHF www.egger.com/downloads/bildarchiv/234000/1_234260_FY_EGGER_DHF_2_pages_EN.pdf [viewed 12.3.16]

5 WHO (2006) Temperature sensitivity of vaccines. http://apps.who.int/iris/bitstream/10665/69387/1/WHO_IVB_06.10_eng.pdf [viewed 12.3.16]

6 WHO, Frequent exposure to suboptimal temperatures in vaccine cold-chain system in India: results of temperature monitoring in 10 states www.who.int/bulletin/volumes/91/12/13-119974/en/ [viewed 12.3.16]

7 Kalkowski J. (2012) Temperature-sensitive vaccines get woollen jackets www.packagingdigest.com/pharmaceutical-packaging/temperature-sensitive-vaccines-get-woolen-jackets [viewed 12.3.16]

8 Stanwix, W. and Sparrow, A. (2014) *The Hempcrete Book*, Green Books

9 Cheshire Oaks (2013) Building Performance Evaluation www.greencoreconstruction.co.uk/downloads/MS_Cheshire_Oaks_Building_Performance_Evaluation_-_Summary_-_Technical_a_.pdf [viewed 12.3.16]

10 Natural Insulations, Thermofolc www.naturalinsulations.co.uk/product/thermofloc/ [viewed 12.3.16]

11 RX3 2009 Review of Market Potential for Cellulose Insulation Products in Ireland (2009) www.rx3.ie/Markets-for-Cellulose-Insulation [viewed 12.3.16]

12 Earth Building UK and Ireland www.ebuk.uk.com/ [viewed 12.3.16]
13 Morton, T. (2008) *Earth Masonry Design and Construction Guidelines*, BRE Press
14 www.selfbuildportal.org.uk/straw-baling-perthshire [viewed 12.3.16]
15 www.self-build.co.uk/natural-building-straw-bale [viewed 12.3.16]
16 www.modcell.com/ [viewed 12.3.16]
17 Riversong HouseWright https://riversonghousewright.wordpress.com/about/9-hygro-thermal/ [viewed 12.3.16]
18 Zero Carbon Hub (2014) *Closing the Gap between Design & As-built Performance: Evidence Review Report*, Zero Carbon Hub
19 www.building.co.uk/how-to-be-cool-tackling-overheating-in-uk-homes/5068037.article [viewed 12.3.16]
20 Sharman, A. and Pickard, J. (2014) Lord Prescott's £60,000 homes are rotting, *Financial Times*, May 16 www.ft.com/cms/s/0/d163240a-dc3f-11e3-8511-00144feabdc0.html [viewed 12.3.16]
21 Taylor Wimpey launched a £5 million claim against the architects www.construction-manager.co.uk/news/rogers-faces-5m-w3rit-o1ver-troub7led-oxley-w9oods/ [viewed 12.3.16]
22 What went wrong at Oxley Woods?, *The Architect's Journal* www.architectsjournal.co.uk/news/what-went-wrong-at-oxley-woods/8662623.article [viewed 12.3.16]
23 Pioneering flats 'could blow down', *The Guardian* www.theguardian.com/society/2005/nov/01/communities.politics [viewed 12.3.16]
24 Zolfagharifard, E. (2014) Are energy efficient homes making us ILL? Toxic mould caused by poor air circulation could trigger 'sick building syndrome', *Daily mail* www.dailymail.co.uk/sciencetech/article-2562146/Are-energy-efficient-homes-making-ILL-Toxic-mould-caused-poor-air-circulation-trigger-sick-building-syndrome.html [viewed 12.3.16]
25 Shrubsole, C., Macmillan, A., Davies, M. and May, N. (2014) 100 Unintended consequences of policies to improve the energy efficiency of the UK housing stock, *Indoor and Built Environment* 23 (3), 340–352
26 Stanwix and Sparrow (2014) op cit.
27 Bristow, C. (2015) *Identifying barriers to the use of natural building materials in mainstream construction in Ireland*, University of Bath, MSC Architectural Engineering
28 http://ww7.global3digital.com/barratt/reports/ar2013/governance/board-of-directors/roderick-maceachrane.html [viewed 12.3.16]
29 Yates, T. Ferguson, A., Binns, B. and Hartless, R. (2013) *Cellulose-based building materials use, performance and risk*, NHBC Foundation NF55 2014
30 Builders warned of risks with natural materials, The Construction Index www.theconstructionindex.co.uk/news/view/builders-warned-of-risk-of-natural-materials [viewed 12.3.16]
31 Warmcel and Improcel live on with a new owner & manufacturer, CIUR www.ciur.co.uk/news/zobrazit/warmcel-and-improcel-live-on-with-a-new-owner-manufacturer [viewed 12.3.16]
32 www.thermafleece.com [viewed 12.3.16]
33 Geiser, K. (2001) *Materials Matter – Towards a Sustainable Materials Policy*, Massachusetts Institute of Technology

12 Policy issues for healthy buildings

A critical analysis

As discussed in Chapters 10 and 11, market forces and current practice work against the development of healthy buildings, ensuring that the vast majority of buildings are constructed using hazardous petrochemical-based materials. It is hardly surprising that poor IAQ is commonplace in modern buildings if there is little regulation to limit or exclude hazardous materials from buildings. This is unlikely to change unless there are significant policy and legislative shifts and much greater awareness among professionals and the general public about healthy building issues. Much of the work to underpin this has been done, particularly in the European Union, but it is not being applied as effectively as it might be in pioneering countries like Germany and France. In the UK and Ireland, this work is largely being ignored. The current debate as to whether the UK should leave the EU will have been resolved by the time this book has been published. If the UK were to leave the EU and remove much European legislation from the UK statute book, as has been suggested, then what little environmental protection currently exists could be done away with. This would be a frightening prospect. While there are problems with EU policies and rules, some of which are discussed below, on balance, EU environmental legislation has had a positive effect.

As set out in the earlier chapters of this book, there is sufficient evidence to support the idea of adopting the precautionary principle and avoiding the use of hazardous materials wherever possible. However there are some that use the excuse that there is insufficient evidence of the health hazards from VOCs, carcinogens and other emissions in buildings. This head-in-the-sand approach is explored in this chapter, as we need to understand why so many people want to believe that chemicals and technology will not harm us. Even many green advocates of energy efficiency take this view. As a result, policy and regulations by official and professional bodies fail to address issues of IAQ and hazardous materials.

For those looking for policy guidance in this chapter, sadly much of it is devoted to a critique of totally inadequate policies and the need for proper policies to be introduced. It is important to try to understand why adequate policies and regulations do not exist.

It is frequently argued that it will take a lot of work to establish standards and certification to control hazardous materials and approve healthy ones, as though this is somehow justification for not making the effort. However it is simply not true that standards and certification do not yet exist. Much of the work has already been done within the EU, with a range of directives, policies and evidence-based standards that are already in place. EU directives and policies are discussed in this chapter. There is also a considerable body of science that has documented the dangers from VOCs, carcinogenic, asthmagenic and mutagenic materials, as has been indicated in earlier chapters.

The biggest problem is the lack of a direct causative link between toxic and hazardous materials and poor health. Much of the evidence is circumstantial and insufficient epidemiological studies have been done to establish clear causative links. Problems with the medical community are also discussed here and questions asked such as why environmental variables are repeatedly excluded from epidemiological studies. Good science should explore all aspects of a problem but so many medical researchers have already made up their minds (or these have been made up for them by the drug companies who fund so much research) as to the hereditary or lifestyle causes of ill health to the exclusion of environmental explanations.

A further problem is the attitude or lack of awareness of the general public and the politicians that they elect. Unless politicians become more alert to the risks they will not be able to resist pressure from the chemical industry lobby groups, advertisers and lobby groups that promote petrochemical and hazardous products.

Then there are problems with the construction industry and the professionals who commission, design and build buildings. Here the level of ignorance is quite shocking, with head-in-the-sand approaches developed to a fine art. There is great resistance to the adoption of unfamiliar and innovative materials and if the supply chain is not robust enough to supply healthy materials, then more hazardous materials will be used. Builders merchants and distributors have little interest in promoting healthy alternatives if they can make sufficient profit from mainstream products and materials. Finally it is necessary to review legislation, such as building and health and safety regulations, and consider what needs to be done to protect workers and building occupants from being surrounded by toxic materials.

Healthy building deniers or avoiders?

We have become familiar with the issue of climate change denial but the debate about whether climate change is happening can distract our attention from other equally important environmental issues. The climate change industry preaches doom and gloom about the future of the planet, but shows surprisingly little interest in the solutions and mitigation measures that might reduce the problems. Positive measures to reduce the use of hazardous materials will have the added benefit of reducing pollution, emissions and damage to ecosystems; so adopting healthy building solutions is a win-win strategy. Surprisingly the importance of healthy buildings and IAQ has been recognised by a UK government climate change report on the health effects of climate change. ‘There is experimental evidence that certain pollutant concentrations in houses, for example those of formaldehyde, may exceed (e.g. due to indoor use of particle boards and adhesives) outdoor air pollution levels’ [1]

Despite this insightful report, a problem exists with healthy building denial or avoidance. Official policies have focused on external air pollution and tried to play down the importance of indoor air pollution. Furthermore there has been a considerable amount of work devoted to ‘wellness’ and town planning policies that tend to act as a smokescreen, obscuring the failure to address the need for healthy buildings. An example of this can be found with the UK Business Council for Sustainable Development (UKBCSD).

> UKBCSD feels that the most important question to how will the wider benefits such as the proven health and financial benefits resulting from good use of technology, design, open space – which in turn have both economic and social benefit (e.g. less impact on other public services, etc.) – be achieved or sustained.[2]

It is welcome that the UKBCSD appears to promote healthy buildings through a healthy buildings focus group. It held a conference on ‘Healthy buildings’ on 9 July 2015 in

Bedfordshire. However despite scouring through 160 pages of powerpoint slides from the event, it was impossible to find anything more than a passing reference to healthy buildings or any of the issues discussed in this book, despite the title of the conference! The chief executive officer of the UKBCSD mentions 'healthy product selection' but gave no details. Another speaker, Rich Frances from the Feeling Good Foundation, managed to ignore healthy buildings apart from a passing reference to a World Green Building Council report discussed below. Jason Longhurst from the hosts, Central Bedfordshire Council, equated healthy buildings to growing vegetables, walking, cycling and fitness trails, mentioning nothing about buildings. Lewis Knight from Bicester Eco Town in Oxfordshire described the development of housing in Bicester but gave no details other than aiming for code 5 energy efficient houses in Elmsbrook; apparently no standards for IAQ have been considered. Bike storage, green space, walking and cycling, play areas, allotments and community orchards were mentioned but nothing about how the buildings will be healthy. However another speaker, a house-builder, Chris Carr, chair of the Federation of Master Builders, did list a healthy home as including good IAQ but gave no details at all, other than to say that houses should be built to a very high standard; and so it went on.

This event is used here as an example of where an organisation had the well-intentioned desire to discuss healthy buildings but only managed to wheel out speakers who were unable to address the subject! This is typical of countless other events held in the UK in the last few years that use the words 'healthy buildings', only to discuss in very vague terms the value of fitness trails and outdoor exercise. The UKBCSD may not have deliberately intended to deny or avoid the importance of healthy buildings, but it like many other organisations, does not know seem to know what they are![3]

There is, of course, nothing wrong with encouraging fitness trails and healthier lifestyles, better diets and a fresh approach to healthcare but this seems to be part of an attempt to define healthy buildings as something to do with general health and wellbeing, excluding the issue of the buildings themselves!

This redefinition of healthy buildings as 'well-being' has been underpinned by academic research at places like Warwick University and Oxford Brookes University. Warwick University established WISE (the Wellbeing In Sustainable Environments) research unit. Dr Libby Burton at WISE, who sadly died in 2014, lectured on 'the pursuit of happiness' and developed a tool to promote design for wellbeing funded by a UK EPSRC Dream Fellowship. Burton also worked on research about getting people outdoors (Inclusive Design for Getting Outdoors). She proposed a kite mark for houses that would meet well-being standards and also worked on dementia-friendly neighbourhoods and care home design. Burton was one of the editors of a massive 736-page reference guide to *Wellbeing and the Environment*, which is part of a six-volume guide to well-being. It includes one chapter by Croxford, which mentions IAQ. However IAQ and healthy buildings remain elusive concepts in this field of academic endeavour where well-being is largely associated with urban design. If the kite mark had been developed then it could have included IAQ standards but this was not what had been envisaged.[4]

Healthy towns

The UK government has picked up on the urban design route to well-being with an initiative to create 'healthy towns'. The Healthy Towns programme invited expressions of interest to 'develop new towns, neighbourhoods and communities to promote health and wellbeing' in 2015 and this attracted 114 proposals, mostly from local authorities in England.[5]

Ten healthy new towns were selected in March 2016, from Hampshire to Darlington, to be supported by Public Health England. The proposal is to design out 'obesogenic' environments and 'design in health and wellbeing'. Schemes are expected to include creating routes for people to walk or cycle, fast-food-free zones and dementia friendly streets.

> Professor Kevin Fenton, National Director for Health and Wellbeing at Public Health England, said: 'Some of the UK's most pressing health challenges – such as obesity, mental health issues, physical inactivity and the needs of an ageing population – can all be influenced by the quality of our built and natural environment. The considerate design of spaces and places is critical to promote good health'.[6]

None of the proposals included any mention of IAQ or healthy buildings and when contacted, none of the ten selected 'new towns' currently have IAQ or healthy buildings policies or strategies to ensure good air quality in the houses that will be built, as this is not on the well-being agenda.

Some have linked indoor environment with well-being through 'active design'. Active design is meant to enable an active lifestyle and this is promoted through the Mobility Mood Place team at Edinburgh and York Universities.[7]

An issue of *Building Research and Information* has discussed academic work on physical activity, sedentary behaviour and the indoor built environment.[8]

The idea of designing towns and places that will lead to wellness is a revival of a fallacy that design can determine the behaviour of people. This was hotly debated some years ago around the issue of 'defensible space' and designing out crime.[9] The myth was put forward that housing layouts could be designed to reduce or eliminate crime and this led to officially backed policies of 'secured by design'. Police resources were devoted to advising people how they could design out crime. What is remarkable is that this lunatic idea gained so much credibility, though thankfully little is heard of it today. There are many excellent critiques of the 'problematic assumptions about the interaction of human behaviour and the environment'; human behaviour may be influenced by design, but it is unable to determine it.[10] Wellness place-making is an attempt to revive false notions of environmental determinism and is likely to be a passing fad, soon forgotten, but in the meantime resources are directed away from tackling the real problems of IAQ.

Why should so much government research funding go to universities to tell us that we shouldn't slump in front of the TV but run up and down stairs (unless we live in a bungalow) or go out for healthy walks in the fresh air. Is it really necessary to commission expensive academic research to tell us this? There are some 15 or more centres of research around this topic in the UK and many hundreds more worldwide. While no doubt well intentioned, this has become an effective tool for government to maintain its head in the sand about hazardous chemicals in buildings. The UK government even runs a National Wellbeing Programme, which measures 'how we are doing' as a nation.[11] There is a Network of Wellbeing (NOW) that promotes 'Action for Happiness', 'natural capital and regenerative capitalism'. But its network does not extend to considering healthy buildings![12] Dulux paints even sell a matt emulsion green paint called 'Wellbeing'.[13] These and other similar initiatives are part of an ideological movement to promote 'wellness and mindfulness' as a diversion from facing up to real world problems.

> The modern idea of wellness is opposed to deep thinking. Instead it encourages us all to become happily stupid athletes of capitalist productivity … the ideology of wellness

> shares with the controversial movement in psychology called 'positive thinking', the twin assumptions that: a) you can be whatever you want to be; and therefore b) if anything bad happens to you, it's no one's fault but your own. In this way, the apparent optimism of the public encouragement to 'wellness' hides a brutal, libertarian lack of compassion. No wonder that, as the authors remind us, Prime Minister David Cameron was so keen to apply positive psychology to the task of measuring the country's happiness.[14]

A critique of the misuse of health and well-being, focused on workplace health issues, comes to similar conclusions.

> Well-being has become one of the most over-used phrases in the English language. Many physical problems that manifest themselves as back pain or RSI have a psychological cause and are as much to do with work-related stress as bad manual handling or repetitive movements. This has led some to conclude that it is more important to concentrate on psychosocial factors at work than physical ones. Unfortunately, some people have misinterpreted this as meaning that you should forget about improving the physical environment and instead try to prevent musculoskeletal disorders through preventing stress at work – a view that could have very dangerous consequences. You have to do both.[15]

'Unfortunately many employers seem to be focusing, *not on keeping workers safe*, but instead trying to encourage them to look after their own health by encouraging them to eat well and exercise'.[16]

Wellbeing in buildings

The World Green Building Council (WGBC) has given some attention to the importance of healthy buildings in a recent report, though this also promotes the *wellbeing* concept:[17] 'Productivity improvements of 8–11% are not uncommon as a result of better air quality'. Sponsored by Lend Lease, Skanska and real estate management company JLL, this report does refer to IAQ but is mainly concerned to convince businesses and commercial building owners that workers will work harder and be more productive if they work in a building designed for health and well-being. Clearly if fewer workers are off sick and greater productivity results from better buildings, that has to be positive and the WGBC report does include a brief section on IAQ, but fails to provide any real guidance on how to deal with this.

> Office occupants can be exposed to a range of airborne pollutants that typically include chemicals, micro-organisms and particles originating from sources both within and outside the building. Ozone, off gassed volatile organic compounds (VOCs), allergens and asthmagens make for a veritable cocktail of potential pollutants that may come from building materials, carpets, finishes, cleaning products, office equipment and traffic … The various health implications associated with poor indoor air quality – from respiratory problems to infections to irritants have been the subject of research for a long time and are well established. But as we better understand the impacts, so our understanding changes on what is considered acceptable or desirable. Design strategies that ensure good air quality are a pre-requisite for a healthy and productive working environment. Although both are important, there is distinction between ensuring a supply of fresh air

> through ventilation, and stopping pollutants at source by minimising the 'off gassing' of materials … There are long established links between building materials and human health, from the formaldehyde found in particle board to the asthmagens found in some kinds of paint, flooring and interior finishes. Fortunately, products such as low and no-VOC paint and green certified furnishings and other fit-out components are available in many markets. Environmental Product Declarations are an important element of transparency. The well-considered selection of healthy finishes and furnishings goes hand-in-hand with adequate ventilation rates to ensure good indoor air quality and should be implemented as part of any sustainable fit-out and ongoing purchasing program.[18]

The WGBC report is a step forward in that it mentions the importance of emissions from materials, but it does not go far enough in explaining how to deal with this or calling for the establishment of limiting standards for emissions. Environmental Product Declarations (EPDs) are mentioned but the WGBC does not point out that these do not include any information on health effects. The report does not spell out that conventional buildings create IAQ problems and this became clear in a presentation by Pottage of Skanska at the Alliance for Sustainable Building Products (ASBP) conference on 'Healthy Buildings: The Role of Products', in January 2016. Pottage, Skanska UK's healthy buildings officer, who was involved in the WGBC report, gave the example of Neelands House, a new office for Skanska in Doncaster, but failed to provide one shred of evidence that the building, whilst energy efficient, was either green or healthy. In fact the only green claim for the building, constructed of prefabricated concrete sandwich panels, was that it used linoleum flooring.[19]

The UK Green Building Council (UKGBC) provides very little detail on its website about healthy buildings, though it does draw attention to the WGBC report and did launch a project on health, well-being and productivity in retail in March 2015 and ran a master class on how to deliver healthy buildings in December 2015, which was to help participants 'understand the definitions of a healthy building and the frameworks and criteria available to assess buildings and their impact on occupants'.[20] [21]

For many years there has been a struggle with the problem of *greenwash* in the construction industry, where conventional products and materials were rebranded as green and eco. Perhaps a new term of '*wellwashing*' will also be needed to explain attempts to gloss over the real problems of IAQ and healthy buildings.

The medical profession: healthy building deniers?

A major barrier to healthier building is the medical profession. Market research by Mcgraw Hill in the USA says that only 32 per cent of general practitioners believe that buildings have an impact on health, though interestingly over 50 per cent of pediatricians do.

> The report, 'The Drive Toward Healthier Buildings: The Market Drivers and Impact of Building Design on Occupant Health, Well-Being and Productivity,' finds that though 18% of homeowners say that doctors are their primary source for information on healthy home products and decisions, only 53% of pediatricians, 32% of family doctors/general practitioners and 40% of psychiatrists believe that buildings even impact patient health. Only 15% report receiving any information on this connection, but the results also reveal that a key challenge is not just getting information to them but gaining their attention in ways that would alter their perspective, with nearly a quarter (22%) reporting that more information would likely not change what they do today.[22]

This confirms that IAQ is not given a high priority in the medical profession as a serious health problem. This is justified by stating that there is insufficient evidence of a causal link between buildings and illness. The lack of evidence is hardly surprising when major epidemiological studies ignore these issues and possible correlations are discounted because data on building conditions are not collected. Research on asthma, COPD and other health problems tends to concentrate on symptoms and even when building conditions, such as the presence of mould and damp are considered, data are usually based on self-reporting by occupants, thus relying on observations by untrained people. Medical researchers rarely visit homes and buildings to take readings of IAQ or inspect damp and mould, and thus it is hardly surprising that evidence is lacking.

An example of this is a major EU health study, the HELIX project (The human early-life exposome – novel tools for integrating early-life environmental exposures and child health across Europe), given the task of examining environmental exposures of chemicals to babies in early development. It involves over 20 medical and university agencies throughout Europe, part of the work costing over EU€38 million, funded under the EU Framework 7 programme.

> In six existing birth cohort studies in Europe, HELIX will estimate prenatal and postnatal exposure to a broad range of chemical and physical exposures. Exposure models will be developed for the full cohorts totaling 32,000 mother–child pairs, and biomarkers will be measured in a subset of 1,200 mother–child pairs.[23,24]

It might have been expected that this study would give serious consideration to indoor air pollution, as there is a substantial body of literature showing that young infants are at particular risk from indoor toxic emissions. The HELIX project even involves installing indoor air monitors in some homes.

> Questionnaires will also collect information on sources of indoor air pollution including ETS, cooking and heating appliances, and ventilation … In the panel studies, indoor air pollution will be measured to characterize errors when using exposure information from questionnaires and models. This will be done using passive samplers for nitrogen dioxide (NO_2) and BTEX (benzene, toluene, ethylene, and xylene), and active PM2.5 (particulate matter with diameter ≤ 2.5 μm) cyclone pumps, installed in the home. The panel studies will measure daily repeat biomarkers of the nonpersistent chemicals (phthalates, phenols, organophosphate pesticides) in urine.[25]

It is worrying that the research did not appear to include tests for significant indoor air hazardous chemicals such as formaldehyde and styrenes, even though these pose the most significant risks to the development of babies and young infants. The HELIX researchers were contacted and they explained that they will be asking occupants about mould and damp, but once again relying on untrained observation. However they will also examine bi-markers of flame retardants.[26] It is possible that there will be some useful data from this study, but even if indoor pollution is identified as a possible cause of pre-natal exposure, the researchers will have no data on what caused this, as they will know nothing about the homes and the materials and construction of the buildings included in the study. Despite an impressive array of medical research bodies involved, they should have included more experts on IAQ and buildings.

Another major medical science research consortium known as Exposomics claims to be carrying out indoor and outdoor monitoring using mobile personal exposure monitoring, however it and one of its partners, Imperial College London, defines indoor air pollution as 'The result of cooking and heating households using solid fuels (i.e. wood, charcoal, coal, dung, crop wastes) on open fires or traditional stoves. In poorly ventilated dwellings, smoke in and around the home can exceed acceptable levels for fine particles 100-fold'.[27]

This doesn't make sense when they are studying the medical effects of pollution in Europe, as cooking indoors with charcoal is a developing country problem. Exposome is ignoring IAQ and appears to be largely focusing on external air pollution. Just as wellness and wellbeing are a smokescreen, so too is the emphasis on external air pollution.[28]

The focus on external air quality at the expense of indoor air problems

Almost every authority from WHO to environmental protection bodies accepts and states clearly that indoor air problems are far more significant in terms of having a damaging effect on health than external air pollution. Yet national governments, medical research consortia and even the EU pour resources and generate media attention on an almost a daily basis about the threat of external air pollution while indoor air problems are side-lined.

The recent Volkswagen diesel car emissions scandal triggered off hundreds of new media reports about the risks of toxic emissions from diesel cars and numerous commentators have suggested that health problems have increased because of this. Hard evidence of this is rarely presented however.

> Last month, the EU announced it was taking legal action against the UK because it was persistently over the safe limit for air pollution – in particular, levels of nitrogen dioxide.
>
> But the UK is not alone – most EU countries struggle to meet the targets, says Martin Adams of the European Environment Agency. In fact, the EU is currently taking action against 17 out of 28 member states with serious air quality problems. Frank Kelly, Professor of Environmental Health at King's College London, puts the blame for the UK's pollution woes squarely on the shift to diesel vehicles over the past 10 years. About 80% of deaths related to outdoor pollution are linked to heart disease and strokes, while 14% are due to lung or respiratory diseases, and 6% to cancer, estimates the WHO.[29]

Reducing pollution from vehicle emissions and other sources would be a good thing, particularly because some of this pollution is sucked into buildings, where it becomes concentrated and adsorbed into materials. However the percentages quoted by the BBC above are meaningless unless they are set in a wider context of epidemiological studies. In practice it is almost impossible to separate out ill health from indoor pollution from external sources, as we are all going in and out of buildings all of the time. But why do these reports never mention IAQ?

The BBC, quoting Professor Kelly, gives the impression that deaths from air pollution are related to external pollution but Professor Kelly is referring to a WHO report that is about deaths from air pollution in general, the majority of which are from indoor air problems. However he claims that this is largely due to biomass cooking in Asia.[30] Professor Kelly is chair of the UK government-sponsored Committee on the Medical Effects of Air Pollutants (COMEAP). COMEAP has been instructed by the government to effectively ignore IAQ for the foreseeable future and concentrate on external air pollution.

> It was agreed that the Committee's work had out of necessity, been dominated by ambient air pollutants. This was due to pressing needs from policy leads, in particular

> from Defra, for consideration of the effects of ambient air pollutants on health to inform their air quality strategy: a health-based strategy. This work has taken priority, despite COMEAP's remit including both ambient and indoor air pollution. Given the current work priorities for COMEAP over the coming years, it is unlikely that COMEAP will be engaging in work on indoor air pollutants, certainly over the coming year, due to similar pressing needs from Defra.[31]

Thus Kelly is making it quite clear to professionals and scientists concerned with indoor air problems, that the UK government has no interest in the issue for the next few years. This is based, not on careful scientific analysis, but political expediency.

> Members will be familiar with the large European Study of Cohorts for Air Pollution Effects (ESCAPE) study funded by the European Commission in which different combinations of approximately 30 European cohorts were included in analyses of associations of air pollutants with different health endpoints. Associations with residential PM2.5, PM2.5 absorbance, PM10, PMcoarse, NO_2, NOx, traffic intensity on the nearest road and traffic load on major roads within 100m were examined.[32]

DEFRA's website contains lots of information on air pollution but the word indoor does not appear once and the DEFRA *Guide to UK Air Pollution Information Resources* published in June 2014 has nothing in it about IAQ.[33]

Some scientists have set about inducing panic among politicians and policy-makers by comparing current external air pollution problems with the 'Great Smog' of 1952.

> It is estimated that around 4,000 people died as a result of the Great Smog of London in 1952. That led to the introduction of the Clean Air Act in 1956. In 2008, 4,000 people died in London from air pollution and 30,000 died across the whole of the UK. The Government needs to act now, as Government did in the 1950s, to save the health of the nation.[34]

In a 79-page UK Parliament environmental audit report, the word 'indoor' does not appear once. Thus it is likely that members of Parliament remain blissfully unaware of the problem of IAQ. Concentrating on external air pollution is politically more expedient as it is much easier to blame truck owners and haulage companies than the manufacturers of hazardous materials that cause indoor air problems.

Healthcare buildings in UK are not necessarily healthy

It might be expected that healthcare buildings would be designed to ensure good IAQ but current UK government procurement rules for healthcare buildings do not require this, apart from meeting a basic BREEAM standard. Official guidance about material selection is only concerned with appearance and cleanliness. As a result it is common for glued uPVC floor finishes to be widely used in healthcare buildings, without thought for the potential emissions within the building.

> Healthcare facilities should provide a therapeutic environment in which the overall design of the building contributes to the process of healing and reduces the risk of healthcare-associated infections rather than simply being a place where treatment takes place.[35]

Perhaps higher standards might have been found from SHINE, a non-profit organisation established to promote exemplary sustainable health care buildings.[36] SHINE has carried out a research project, 'Making Existing Healthcare Buildings Sustainable', which focused on energy and CO_2 emissions, benefits to health and patient recovery and other savings. However nothing about IAQ was apparent on the SHINE website or in any other UK guidance documents on healthcare and hospital buildings. SHINE case studies describe energy efficiency and airtight buildings with exposed concrete walls and ceilings but no awareness of the health hazards of different materials or forms of construction. Perhaps this would be acceptable if IAQ was not a key factor in healthcare buildings but there has been considerable research, internationally, into IAQ in hospitals dealing with hazardous emissions, mould and fungus and different approaches to ventilation. 'The Center for Disease Control and Prevention estimates that more than 2 million patients a year acquire infections in U.S. hospitals while they are hospitalized for other health problems, and that 88,000 die as a direct or indirect result'.[37 38 39]

Many hospitals are becoming increasingly specialised to deal with specific problems, such as the Lane Fox REMEO Centre at East Surrey Hospital, provided by the BOC and Linde Healthcare group, which is an inpatient unit for patients with complex respiratory problems.

> The centre also provides inpatient care for those failing to 'wean' from invasive ventilation following critical illness, and investigates chronic respiratory failure resulting from a variety of causes such as:
>
> - chronic obstructive pulmonary disease (COPD)
> - neuromuscular disease (for example, Duchenne muscular dystrophy and motor neuron disease)
> - chest wall disease (for example, scoliosis)
> - chronic respiratory failure related to obesity.[40]

Despite the innovative nature of this building, IAQ issues did not appear to have been considered and there had not been any IAQ tests carried out, but the architects have expressed an interest in taking this into account in similar future projects.

Other specialised treatment buildings and hospices dealing with cancer and asthma also fail to consider the use of asthmagenic and carcinogenic materials in their buildings. To date much of the work on air quality in healthcare buildings has concentrated on ventilation measures but possibly with the introduction of IAQ assessment into BREEAM more attention will be paid to this. 'The real option value of the economic benefits of better IAQ is almost 4 million euros and the real option pay-off of the IAQ investment exceeds 0.5 million euros'.[41]

EU policies on IAQ and materials

Work on eco-labelling of construction products has been under debate in Europe for many years. There are numerous eco-labels and certification systems but the most common today are EPDs. The EU issued its Construction Products Directive in 1989 and this eventually led to the Construction Product Regulations (CPR) in 2011, which came into force in 2013.[42] The aim was to create harmonised regulations for standards of materials that would also ensure sustainability and responsible resource use. Compliance with the CPR allows product manufacturers to add a CE label to their products. There has been a continual process of developing standards particularly in relation to sustainability through a working group called CEN/TC350.[43,44] However mainstream construction industry vested interests have been very effective in minimising the impact of CPR, and this has made it relatively easy to comply with the sustainability requirements. Only Germany and its Institut Bauen and Umwelt enthusiastically adopt the full intentions of the CPR, particularly in relation to sustainability.[45]

What is important about CPR for this book was the addition of BRCW 3 (Hygiene, health and environment).

> BR (Basic Requirements for Construction Works)
> 3 Hygiene, health and environment
> The construction works must be designed and built in such a way that they will, throughout their life cycle, not to be a threat to hygiene and safety of workers, occupants or neighbours, nor have an exceedingly high impact, over their entire life cycle, on the environmental quality or on the climate during their construction, use and demolition, in particular as a result of any of the following:
>
> (a) the giving-off of toxic gas;
> (b) the emissions of dangerous substances, volatile organic compounds (VOC), greenhouse gases or dangerous particles into indoor or outdoor air;
> (c) the emissions of dangerous radiation;
> (d) the release of dangerous substances into ground water, marine waters, surface waters or soil;
> (e) the release of dangerous substances into drinking water or substances which have an otherwise negative impact on drinking water;
> (f) faulty discharge of waste water, emission of flue gases or faulty disposal of solid or liquid waste;
> (g) dampness in parts of the construction works or on surfaces within the construction works.[46]

The Basic Requirements for Construction Works are very comprehensive and, if enforced, would have the effect of reducing the use of hazardous materials in buildings. What is curious about the wording of the EU regulations is the use of the term '*construction works*' rather than construction materials or products, as the regulation is concerned with 'harmonised conditions for the marketing of construction *products*'. This provides the manufacturers with a loophole to avoid complying with Annex 1 basic requirements as they refer to works not products. Clearly to meet these standards, products and materials should not give off toxic gases or emit VOCs, etc. however product declarations issued by manufacturers avoid the issue by stating 'No Performance Declared'.

Air permeability	NPD No performance declared
Water impermeability	NPD
Release of dangerous substances	Declaration filed with the Technical Approval Body
	ETA 13/0286

Figure 12.1 Typical extract from a product declaration BR3 (hygiene, health and environment)

Thus product declarations are almost useless as a source of information about a product as a result of both the CPR and the EU REACH directive failing to ensure that hazardous problems are openly declared. Having introduced stringent environmental protection rules, the EU has made it possible for manufacturers to avoid telling us anything about their products! The following extract shows how the EU has confused the issue.

> Next to the harmonised standards, in future there will be no European Technical Approvals based on agreed guidelines or assessment criteria, instead there will be the European Assessment Documents (EADs) serving as harmonised specifications. They serve as a basis for issuing a European Technical Assessment, in which information about a product's performances are established. The report cannot deliver specific guidelines on how the new sustainability requirement can be implemented, because no side has yet provided these. To date neither the Commission nor any one or several Member States have stated any respective guidelines. However, this may actually present an opportunity, because the lack of such guidelines leaves room for ideas and procedures on how best this new requirement could be implemented. The report includes suggestions by the authors considering what requirements a construction product should meet to be able to be referred to as sustainable.[47]

There was an assumption on the part of the EU that regulatory provisions are already in place to ensure limited emissions from construction products, even though assessment methods did not really exist and the Commission realised that they had to set up measurement and test standards for the regulation of dangerous substances. CEN's working group, CEN/CENELEC TC/351, was requested to develop assessment methods.[48] The following documents have been issued concerned with this.

> CEN Technical Committee TC/351 has undertaken the work requested by Mandate M/366. In January 2014 the Technical Committee finalised the following documents:
>
> CEN/TS 16516:2013: Construction products – Assessment of release of dangerous substances – Determination of emissions into indoor air.
>
> CEN/TR 16496:2013: Construction products – Assessment of release of dangerous substances – Use of harmonised horizontal assessment methods.
>
> CEN/TR 16410:2012: Construction products – Assessment of release of dangerous substances – Barriers to use – Extension to CEN/TR 15855 Barriers to trade.
>
> CEN/TR 16220:2011: Construction products – Assessment of release of dangerous substances – Complement to sampling.
>
> CEN/TR 16098:2010: Construction products – Assessment of release of dangerous substances – Concept of horizontal testing procedures in support of requirements under the CPD.
>
> CEN/TR 16045:2010: Construction products – Assessment of release of dangerous substances – Content of regulated dangerous substances – Selection of analytical methods.

> CEN/TR 15858:2009: Construction products – Assessment of the release of regulated dangerous substances from construction products based on the WT/WFT procedures.
>
> CEN/TR 15855:2009: Construction products – Assessment of release of dangerous substances – Barriers to trade.[49]

The Commission employed a 'contractor' to review a number of questions, such as whether a complete list of substances was provided within product declarations and whether:

> there has been a reduction of hazardous substances used and fewer cases of illness as a result. To respect this obligation the Commission has procured the independent 'Study on specific needs for information on the content of dangerous substances in construction products' with the overarching objective to clarify and analyse the existence of specific needs for information on content of construction products.[50]

A link to the report of this study has been deleted from the Europa website and so the results of this review could not be found. It is very difficult to obtain any accurate information of the impact of the CPR in relation to the emission of hazardous substances, etc. However, the industry is meant to be working towards the declaration of VOC emissions from products as part of the CE marking. Much discussion has taken place around agreed testing methods to establish VOC emissions levels.[51,52]

The CEN website is a poor source of information about its work. The TC 351 group is chaired by Jeroen Bartels, a Dutch environmental management expert.[53] It lists 'involved parties' in its work, which includes a wide range of industry bodies such as PU Europe, the Federation of Rigid Polyurethane Foam Associations, the European Federation for Construction Chemicals, the European Insulation Manufacturers Association and the European Wood Preservative Manufacturers, as well as government standards bodies from many countries.[54]

It is possible to access some summary minutes of TC 351 meetings and plenary sessions and it is clear from these minutes that progress towards widely accepted standards for emissions assessment and testing is slow.

Eurofins, a commercial testing body, provides a useful source of information into the complex and confusing world of assessment of VOC emissions from construction products.[55] It explains that there are different values used by several different emission rating schemes as 'limit values' for evaluating VOC emissions: LCI, CLI, NIK and CREL.[56]

LCI values seem to be the most widely accepted and a European Commission 'sub-group' has been meeting to 'derive and recommend EU-wide harmonized health based reference values for the assessment of product emissions.

> We were established by EC-JRC, Ispra in 2011 (Ispra is a small Italian town with a nice hotel overlooking Lake Maggiore) and are now constituted as a Sub-group of the Expert Group on Dangerous Substances within the larger committee structure of the EC Advisory Group on Construction Products. We use set procedures and protocols to establish EU-LCI values in a rigorous, scientific and transparent manner.[57]

Chaired by Derrick Crump from Cranfield University in the UK, the LCI group is a mixture of academics, government agency representatives, the European Chemical Industry Council and BASF, the multinational chemical company. The status of EU-LCI values is not at all clear and so far the group has only considered VOCs. LCI values were published in a master list in December 2015. 'Very volatile organic compounds (VVOCs), semi-volatile organic compounds (SVOCs) and carcinogens have not yet been considered at this stage'.[58]

LCIs are health-based values of emissions after 28 days from a single product. They are expressed as ug/m^3. These values are guidelines, part of a flexible framework and are not meant to be standards for IAQ but to assess emissions from individual materials and products. It is unlikely that the LCI concept would mean much to those attempting to select low-emission materials, however they are a huge step forward in providing a tool to classify materials and provide accurate information on hazardous emissions from materials and products.

Germany and France have already adopted some different LCI values in standards. (In Germany the organisation responsible is the AgBB and in France AFSSET.)According to Reinhard Oppl at Eurofins, out of a total of 176 compounds, 86 have agreed interim EU-LCI values and a further 90 are still to be harmonised. However there is a long way to go before emissions standards are widely adopted.[59]

> The German Federal Environment Agency (UBA) is campaigning for the European Commission to give its support to the harmonisation of activities in order to ensure that Member States do not concurrently develop any new national evaluation standards and may instead refer to common European LCI values as necessary. It is working to ensure that the Commission facilitates the continuation of these activities so that a harmonised LCI list will be in place and published as soon as possible. Such a list would make a significant contribution to the implementation of legally defined objectives concerning the harmonisation of the methods for the evaluation of building products and bring about a uniform high level of consumer protection against gaseous emissions from building products throughout Europe.[60]

The LCI group lost its funding from the EU in 2013, indicating the low priority attached to this by the EU. It continued under its own steam, though it may be now receiving new support. Without further funding a full list of LCI value for hazardous compounds may not be available for some time.[61]

What is not clear from the available literature about LCIs is the relationship of emission standards for products with both the REACH directive and construction product regulations. The UK government gives almost no guidance on the CPR other than to give links to EU websites. The UK Construction Product Regulations 2013 say nothing about health and emissions.[62]

REACH

Adopted in 2006, the REACH regulation required all chemical manufacturers to justify the consumer safety of their products. However there has been pressure in recent years to review and water down the impact of REACH. Over 100,000 chemicals have to be reviewed by the final deadline of 2018 but only 47 Substances of Very High Concern (SVHCs) have been listed. Chemsec, an environmental lobby group, wants 378 substances, particularly endocrine-disrupting substances such as phthalates, included as SVHCs.[63]

All chemicals listed under REACH have to be registered with the European Chemicals Agency (ECHA) in Helsinki, Finland, and details of the chemicals including toxicity should be listed on their website.[64] However in order to get details of a product it is necessary to have the substance name, a CAS number and the EC/list number. It is no good putting in the trade name of a product, as this will not bring up any details. Manufacturers can hide their product behind the substance name, however it is possible to search for the candidate list for SVHCs. Many substances are listed as having no effect with Derived No Effect Levels (DNELs).[65]

EU-LCI values have been derived in part from hazardous substances listed under REACH but there are concerns about the relationship between LCIs and REACH.[66] A serious and possibly unintended consequence of the REACH directive is that product safety data sheets no longer include much useful information as to hazardous contents of the product if above 0.1 per cent in weight. Whereas in the past, such data sheets would have provided a lot of useful information about the nature and contents of a product, if it is listed on the ECHA database in Helsinki, almost no information is given to the public. Access to the database is possible but very difficult. Thus despite legislation to restrict the use of hazardous materials and products and more sophisticated testing and assessment tools, together with harmonised standards, the system itself is not user friendly for consumers or specifiers, and manufacturers can hide behind the EU red tape and confusion. A useful way around this is to look for the data sheets for identical products on Canadian websites, as these sometimes follow the old UK or European content and include information on hazardous constituents.

Building regulations: UK and Irish policies

The Irish building regulations suggest that it would be a good idea to reduce pollutants at source to improve IAQ but imply that this is really an issue of ventilation standards. There does not appear to be any enforcement of the guidelines on reducing pollutant sources.

> 1.1.6 Airborne pollutants include those that are released from materials and products used in construction, decoration and furnishing of a building, and those created as a result of the activities of the building occupants. Common pollutants in a dwelling are combustion products from unflued appliances (e.g. gas cookers) and chemical emissions from construction and consumer products. In an office building, body odour is the key pollutant but there are a number of other pollutant sources including building materials, furnishings, printers and photocopiers. Where pollutants can be reduced at source through the use of low emission materials and products this will improve indoor air quality.[67]

In the UK building regulations the following standards are set in Part F1.

> Exposure to the following contaminants should not exceed:
>
> - NO_2 288 μ g/m^3 150 ppb-1 hour average
> - 40 μg /m^3 long term average
> - CO 100 mg/m^3 90 ppm 15 min average
> - 60 mg/m^3 50 ppm 30 min average
> - 30 mg/m^3 25 ppm 1 hour average time
> - 10 mg/m^3 10 ppm 8 hours average time
> - TVOC (Total Volatile organic compounds) 300 μg/m^3 average over 8 hours

However meeting these standards is again seen as a ventilation issue. UK and Irish regulations do not currently address LCIs or hazardous emissions from products or lay down any standards for these to be considered.

Public health ignores IAQ

Public Health England carries out research that covers a wide range of factors that are determinants of health. However IAQ, mould and damp are not included at all as far as could be determined.

> The Public Health Outcomes Framework – Healthy lives, healthy people: Improving outcomes and supporting transparency sets out a vision for public health, desired outcomes and the indicators that will help us understand how well public health is being improved and protected.[68]

A search of the Medical Research Council (MRC) website yielded not a single publication or research report on IAQ. The MRC did fund a ten-month networking project called WELLINE (Wellbeing and the Indoor Environment) led by Professor Ayres at Birmingham University, in collaboration with members of the UK Indoor Environments Group. However it did little more than flag up a few issues that were discussed at various seminars and IAQ was mentioned.

> This proposal will mostly focus on how the indoor environment influences chronic disorders affecting the musculoskeletal, cardiopulmonary and nervous systems, amongst the most prevalent conditions found in the older population, we will embrace the concept of age friendliness and the need to be preventive throughout life.
>
> There is a need to evaluate the significance of the impact of the indoor environment on health compared, for example, to the effects of diet and exercise. What specifically is the contribution of (poor) indoor air quality to chronic disorders, and is there a difference in time spent in the home between the chronically ill and older well people?[69]

The National Institute for Health and Care Excellence (NICE) has barely touched on the subject, though the Royal College of Physicians report discussed in Chapter 1 has drawn attention to the issue of IAQ and this may lead to greater interest in the future.

BRE Home Quality Mark

A major step forward to greater consideration of IAQ can be found in the recently introduced BRE Home Quality Mark (HQM). The HQM includes significant IAQ criteria and this is the first time in the UK that this has occurred in any standards. Previous BRE tools for housing such as Eco-Homes and the Code for Sustainable Homes failed to address the health impacts of buildings and materials, though they did include some requirements concerned with 'Health and Well-being' (Category 7). These requirements ignored materials and IAQ, referring only to daylighting, sound insulation, private space and lifetime homes.[70]

For a brief time the Code For Sustainable Homes (developed by the BRE) was a requirement for all social housing projects funded by UK government (some examples mentioned in this book refer to Code Level 4 or 5) and the Welsh Assembly made it a requirement for all new homes. Despite the fact that the Code included some good features, it was abandoned by the Conservative Government as part of its 'bonfire of red tape' in 2015.[71]

BRE has therefore introduced the HQM as a voluntary substitute for the Code. Many voices in the planning and house building industry, however, have called for it not to be introduced as a mandatory standard and questioned whether homeowners would be willing to pay for the extra cost.[72]

Currently it is too early to say whether there will be much take up of the HQM but its inclusion of IAQ in the standard, mirroring its introduction to BREEAM, is to be welcomed. Section 9 of HQM refers to indoor pollutants and includes criteria for minimising emissions from building product types, formaldehyde and TVOCs. Measurements have to be taken after the building is complete but before occupation, using active pumped sampling (as described in Chapter 2) in the main bedroom and living area. The sampling has to be carried out by accredited organisations and laboratories. There are some contradictory aspects to the standard, as fungal resistant paints must be used in wet areas, even though most fungicides are likely to add to poor IAQ. Emission limits are set out in Table 09.02 in the Home Quality Mark guide.

This is a huge step forward in terms of addressing IAQ issues in buildings in the UK but it is hard to believe that many commercial house builders will be willing to meet the costs of the HQM assessment procedure and the IAQ testing. A further flaw in the HQM is that there is almost no guidance on the use of construction materials to ensure good IAQ. Section 18 refers to the rather curious standard of responsible sourcing of materials. Responsible sourcing has ignored the issue of healthy buildings, though in the HQM environmental product declarations appear to be a requirement. Section 19 refers to environmental impact of construction products, based on lifecycle assessment criteria, which relate to CO_2 emissions rather than health impacts. Section 20 deals with lifecycle costing and durability of construction products. Somewhat surprisingly the HQM does not promote the use of the highly flawed *Green Guide to Specification*, which only gets mentioned once in Section 19.

In order to achieve good IAQ and meet the emission levels in the standard, advice should be given on the selection of materials that have low emissions. Builders are likely to continue to use many standard synthetic and petrochemical-based materials and finishes, unaware that this may lead to them failing to meet the IAQ emission standard. A standard that does not adopt robust accreditation for materials such as LCIs, discussed above, or Natureplus certification, will not be sure of achieving good IAQ.

A scan of the 75 accredited assessors for HQM on the BRE website shows that 90 per cent are in London and the southeast of England. It was hoped to get some feedback from assessors about their experience of applying Section 9 on indoor pollutants, but after contacting 20 assessors, not one said they had actually carried out an assessment. It maybe some time before the HQM is taken up by the house building industry.

Much remains to be done to achieve greater understanding and awareness of healthy buildings, IAQ and the construction materials that are such a key feature of this. A lot of work has been done to assess emissions of hazardous chemicals from products and materials and to set a range of emissions standards. However much of this may be well off the radar of the average, architect, specifier and builder or their clients. It is hoped that this book will at least raise awareness of the issues and provide a source of information that will allow others to pursue the possibility of healthier buildings more vigorously. Unless this happens many of us remain at risk of our health being damaged by hazardous materials in buildings. Vague talk of wellness and healthy places will not be enough to protect our health and that of our children in the future without robust action on the materials that we use in our buildings.

Notes

1 Vardoulakis, S. and Heaviside, C. (eds) (2012) *Health Effects of Climate Change in the UK 2012: Current evidence, recommendations and research gaps*. UK Health Protection Agency
2 Business Council for Sustainable Development, Placing people at the centre of healthy buildings http://ukbcsd.org.uk/2015/09/01/placing-people-at-the-centre-of-healthy-buildings/ [viewed 22.3.16]
3 UK BCSD, Healthy buildings Conference, Central Bedfordshire Council, 9 July 2015, presentations
4 Cooper, R., Burton, E. and Cooper, C.L. (eds) (2014) *Wellbeing: A Complete Reference Guide, Volume II, Wellbeing and the Environment*. Wiley-Blackwell
5 National Health Service (2015) *The Forward View into Action: Registering interest to join the healthy new towns programme prospectus*. NHS
6 National Health Service (2016) NHS Chief announces plan to support ten healthy new towns www.england.nhs.uk/2016/03/hlthy-new-towns/ [viewed 22.3.16]
7 Mobility Mood and Place, a Lifelong Health and Wellbeing funded research programme https://sites.eca.ed.ac.uk/mmp/ [viewed 22.3.16]
8 *Building Research & Information*, Physical Activity, Sedentary Behaviour and the Indoor Built Environment, Guest editors: Alexi Marmot and Marcella Ucci. The Bartlett Faculty of the Built Environment, University College London www.tandf.co.uk/journals/cfp/rbricfp3.pdf [viewed 22.3.16]
9 Coleman, A.M. (1985) *Utopia on trial: Vision and reality in planned housing*. Hilary Shipman
10 Steventon, G. (1996) Defensible Space: A critical review of the theory and practice of a crime prevention strategy. *Urban Design International* 1 (3), 235–245
11 The National Archive, Measures of National Well-being http://webarchive.nationalarchives.gov.uk/20160105160709/http://neighbourhood.statistics.gov.uk/htmldocs/dvc146/wrapper.html [viewed 22.3.16]
12 Network of Wellbeing www.networkofwellbeing.org/ [viewed 22.3.16]
13 Dulux, Wellbeing www.dulux.co.uk/en/colour-details/wellbeing [viewed 22.3.16]
14 Cederström, C. and Spicer, A. (2015) *The Wellness Syndrome – exploitation with a smiley face* www.theguardian.com/books/2015/jan/22/the-wellness-syndrome-carl-cederstrom-andre-spicer-persuasive-diagnosis [viewed 22.3.16]
15 TUC (2015) *Work and Well-being A trade union resource* www.tuc.org.uk/workplace-issues/learning-and-training/health-and-safety/workplace-health-safety-and-welfare/work [viewed 22.3.16]
16 Robertson, H. (2015) Hazards Special report, *Hazards Magazine* www.hazards.org/workandhealth/hardtoswallow.htm [viewed 22.3.16]
17 World Green Building Council (undated but issued in 2015) *Health, Wellbeing & Productivity in Offices: The next chapter for green building*. WGBC
18 World Green Building Council (2015) op cit.
19 Pottage email to author 16 March 2016
20 UK Green Building Council, UK Green Building Council, launches new project on health, well-being and productivity in retail www.ukgbc.org/sites/default/files/PRESS%20RELEASE%20-%20UK%20Green%20Building%20Council%20launches%20new%20project%20on%20health,%20wellbeing%20and%20productivity%20in%20retail.pdf [viewed 22.3.16]
21 UK Green Building Council, Masterclass: How to deliver healthy buildings www.ukgbc.org/course/masterclass-how-deliver-healthy-buildings-1302 [viewed 22.3.16]
22 McGraw Hill (2014) *The Drive Towards Healthier Buildings: The Market Drivers and Impact of Building Design on Occupant Health, Well-Being and Productivity*, Smart Market Report

23 The HELIX Project (2014) Tracking exposome in real time, *Environmental Health Perspectives* 122 (6), A 169
24 EU, Research & Innovation Environment http://ec.europa.eu/research/environment/index_en.cfm?pg=health [viewed 22.3.16]
25 Helix, building the early life exposure www.projecthelix.eu [viewed 22.3.16]
26 Email from Oliver Robinson CREAL Barcelona Institute for Global Health, 13 October 2015
27 Grantham Institute – Climate Change and the Environment www.imperial.ac.uk/grantham/our-work/impacts-and-adaptation/health/air-pollution-exposomics/ [viewed 22.3.16]
28 www.exposomicsproject.eu/our-research/air-pollution [viewed 27.3.16]
29 UK air pollution: How bad is it?, BBC News www.bbc.co.uk/news/uk-26851399 [viewed 22.3.16]
30 Personal email from Frank Kelly, 8 October 2015
31 In a letter from Kelly to Dr Marcella Ucci chair, UKIEG, dated 10 June 2015
32 COMEAP/2014/01 COMEAP, Statement on the Evidence for Differential Health Effects of Particulate Matter According to Source or Components
33 About air pollution, Department for Environment Food and rural affairs http://uk-air.defra.gov.uk/air-pollution/ [viewed 22.3.16]
34 House of Commons, Environmental Audit Committee, *Air quality: A follow up report*, Ninth Report of Session 2010–12 Volume I: Report, together with formal minutes, oral and written evidence
35 Department of Health (2014) Health Building Note 00-01 General design guidance for health-care buildings
36 Shine, sharing knowledge http://shine-network.org.uk/ [viewed 22.3.16]
37 Riley, D., Freihaut, J., Bahnfleth, W.P. and Karapatyan, Z. (no date) Indoor Air Quality Management and Infection Control in Health Care Facility Construction, IAQ T3S1 Innovative techniques in IAQ www.engr.psu.edu/iec/publications/papers/indoor_air_quality.pdf [viewed 22.3.16]
38 Dikmen, Ç.B. and Gültekin, A.B. (2011) Sustainable Indoor Air Quality (IAQ) in Hospital Buildings, *Environmental Earth Sciences* 5, 557–565
39 Famaswamy, M., Al-Jahwari, F. and Al-Rajhi, S.M.M., (1969) IAQ in Hospitals – Better Health through IAQ Awareness https://oaktrust.library.tamu.edu/bitstream/handle/1969.1/94139/ESL-IC-10-10-88.pdf?sequence=1&isAllowed=y [viewed 22.3.16]
40 Surrey and Sussex NHS, Lane Fox REMEO Respiratory Centre www.surreyandsussex.nhs.uk/our-services/a-z-of-services/respiratory/lane-fox-remeo-respiratory-centre/ [viewed 22.3.16]
41 Kajander, J.K., Sivunen, M. and Junnila, S. (2014) Valuing Indoor Air Quality Benefits in a Healthcare Construction Project with Real Option Analysis, *Buildings* 4 (4), 785–805; doi:10.3390/buildings4040785
42 Construction Products Directive/Regulation www.constructionproducts.org.uk/sustainability/products/construction-products-directiveregulation/ [viewed 22.3.16]
43 CEN/TC 350 Sustainability of construction works https://standards.cen.eu/dyn/www/f?p=204:7:0::::FSP_ORG_ID:481830&cs=181BD0E0E925FA84EC4B8BCCC284577F8 [viewed 22.3.16]
44 Construction Products Regulation (CPR) http://ec.europa.eu/growth/sectors/construction/product-regulation/index_en.htm [viewed 22.3.16]
45 Sustainability of construction products associations EPDs: A solution? arge.org/download/43/ARGE_Peters_EPDs.pdf [viewed 22.3.16]
46 Regulation (EU) no 305/2011 of the European parliament and of the council of 9 March 2011 laying down harmonised conditions for the marketing of construction products and repealing Council Directive 89/106/EEC Annex 1.3
47 Kirchner, D. (2011) Report ‘European assessment documents for sustainable construction products: Information for SMEs’ within the framework of the Baltic Sea Region ‘SPIN’ Project Deutsches Institut fur Bautechnik Federal Environment Agency

48 European Committee for Standardization www.cen.eu [viewed 22.3.16]
49 European Commission Brussels, (2014) Report from the commission to the European Parliament and the Council as foreseen in Article 67(1) of Regulation (EU) 305/2011 COM511 final 7.8.2014 http://ec.europa.eu/transparency/regdoc/rep/1/2014/EN/1-2014-511-EN-F1-1.Pdf [viewed 27.3.16]
50 European Commission Brussels (2014) op cit.
51 New European VOC Emissions testing methods www.eurofins.com/media/9591101/gst-2014-x738-oppl-cen-ts-16516.pdf [viewed 22.3.16]
52 CEN/TC 351 – Construction Products – Assessment of release of dangerous substances https://standards.cen.eu/dyn/www/f?p=204:7:0::::FSP_ORG_ID:510793&cs=1EA0E9CEA95C7E7A1CEFC33C03BA48833 [viewed 22.3.16]
53 Jeroen Bartels, Linkedin www.linkedin.com/in/jeroenhmbartels [viewed 22.3.16]
54 CEN/TC 351 (EN) Involved parties in CEN/TC351 'Construction Products: Assessment of release of dangerous substances www.nen.nl/Standardization/Join-us/Technical-committees-and-new-subjects/TC-ConstructionBuilding/CENTC-351-EN.htm [viewed 22.3.16]
55 VOC Emissions under Construction Products Regulation www.eurofins.com/product-testing-services/information/compliance-with-law/european-directives-and-laws/construction-products/voc-emissions-under-cpr.aspx [viewed 22.3.16]
56 VOC Emissions into indoor air limit values (LCI, NIK, CLI, CREL) www.eurofins.com/product-testing-services/services/testing/voc-testing/emissions-into-indoor-air-limit-values.aspx [viewed 22.3.16]
57 www.eu-lci.org/EU-LCI_Website/Home.html [viewed 22.3.16]
58 Lists of substances and EU-LCI values www.eu-lci.org/EU-LCI_Website/EU-LCI_Values.html [viewed 22.3.16]
59 LCI Values, Eurofins http://nsf.kavi.com/apps/group_public/download.php/19938/LCI%20summary%20by%20Reinhard%20Oppl%202012-12-18.pdf [viewed 22.3.16]
60 Harmonisation of health assessment – LCI values and EU-LCI values www.umweltbundesamt.de/en/topics/health/environmental-impact-on-people/indoor-air-hygiene/substances-in-building-products/harmonisation-of-health-assessment-lci-values-eu [viewed 22.3.16]
61 EU group reveals assessment system for indoor emissions https://chemicalwatch.com/23318/eu-group-reveals-assessment-system-for-indoor-emissions [viewed 22.3.16]
62 Construction products regulations & CE marking, including UK product contact point for construction www.planningportal.gov.uk/buildingregulations/buildingpolicyandlegislation/cpr [viewed 22.3.16]
63 EU prepares to re-open REACH 'can of worms' www.euractiv.com/climate-environment/eu-prepares-open-reach-worms-news-507129 [viewed 22.3.16]
64 http://echa.europa.eu/ja/information-on-chemicals/registered [viewed 22.3.16]
65 European Chemical Agency http://echa.europa.eu/candidate-list-table [viewed 22.3.16]
66 European collaborative action 2013: urban air, Indoor Environment and Human exposure Environment and Quality of Life Report No 29 Harmonization framework for health based evaluation of indoor emissions from construction products in the European Union using the EU-LCI concept.
67 Building Regulations Ventilation Part F (2009) Technical Guidance Document, Environment Heritage and Local Government
68 Public Health Outcomes Framework www.phoutcomes.info/ [viewed 22.3.16]
69 Welline Final Workshop: Summary The Indoor Environment and Chronic Disorders across the Life Course, Cranfield University, 9–10 February 2010
70 Code for Sustainable Homes Technical Guide November 2010 www.gov.uk/government/uploads/system/uploads/attachment_data/file/5976/code_for_sustainable_homes_techguide.pdf [viewed 22.3.16]
71 www.building.co.uk/code-for-sustainable-homes-scrapped/5074697.article [viewed 22.3.16]
72 New BRE Sustainable Homes Standard www.turley.co.uk/intelligence/new-bre-sustainable-homes-standard [viewed 22.3.16]

Appendix A

Carcinogenic chemicals used in buildings and building materials

Based on the US Department of Health and Human Services National Toxicology Programme *Report on Carcinogens* (13th edition, 2014), other sources and www.psr.org/environment-and-health/confronting-toxics/examples-of-environmental-carcinogens.html, examples of environmental carcinogens have been compiled.

Please note that all the chemicals listed here are considered 'reasonably anticipated' to be a human carcinogen or confirmed as a carcinogen. Where possible reference is made to their use in building materials.

Acetaldehyde is used primarily as a chemical intermediate in the production of acetic acid, pyridine and pyridine bases, peracetic acid, pentaerythritol, butylene glycol, and chloral. Used in phenolic and urea resins.

2-Aminoanthraquinone is used as an intermediate in the industrial synthesis of anthraquinone dyes and pharmaceuticals These dyes are used in automotive paints, high-quality paints and enamels, plastics, rubber, etc.

Arsenic is no longer produced in the US and many uses of arsenical pesticides have been banned and the number of workers exposed to arsenic likely has decreased since the early 1980s. Nevertheless, occupational exposure to arsenic (including forms other than inorganic compounds) is likely in several industries, including nonferrous smelting, wood preservation and glass manufacturing. Arsenic is a naturally occurring element. It is most commonly used as a wood preservative (in pressure-treated wood) and can be found in building materials, industry and water (inorganic) as well as fish and shellfish (organic compounds). Exposure is through inhalation or ingestion (intentional poisoning). Arsenic is linked to lung cancer, skin cancer and urinary tract cancer. Arsenic is a known human carcinogen.

Asbestos is a group of naturally produced chemicals composed of silicon compounds. It is used in insulation materials due to heat resistance. Human exposure is through inhalation (from disruption of materials containing asbestos) and ingestion (contaminated food/water). Tiny asbestos fibres in the air can get trapped and accumulate in the lungs. Asbestos is linked to increased risk of lung cancer and development of mesothelioma (cancer of the thin lining surrounding the lung (pleural membrane) or abdominal cavity (the peritoneum) and laryngeal cancer. Cancer may appear 30 to 50 years after exposure. Asbestos is a known human carcinogen. The four commercially important forms of asbestos have been chrysotile, amosite, anthophyllite and crocidolite, however, commercial use of anthophyllite was discontinued by the 1980s (IPCS 1986, HSDB 2009). Chrysotile, amosite and particularly crocidolite all have extremely high tensile strengths and are used extensively as reinforcers in cements, resins and plastics. By the 1990s, chrysotile accounted for more than 99 per cent of US asbestos consumption. By 2008, chrysotile was the only type of asbestos used in the US, with uses in asbestos cement pipe, flooring and roofing. In 2009 roofing products accounted for 65 per cent of US consumption. Asbestos is still legally produced in many European countries.

Basic red 9 monohydrochloride is used as a biological stain and as a dye for plastics, glass, etc.

Benzene is used primarily as a solvent in the chemical and pharmaceutical industries. As a raw material, it is used in the synthesis of ethylbenzene (used to produce styrene), cumene (used to produce phenol and acetone), cyclohexane, nitrobenzene (used to produce aniline and other chemicals), detergent alkylate (linear alkylbenzene sulfonates) and chlorobenzenes and other products. Benzene is used as a solvent in chemical and pharmaceutical industry, and is released by oil refineries. It is one of the largest-volume petrochemical solvents in production; it is produced from coal and from petroleum. Exposure is through inhalation (smoke, gas emissions, etc.) or ingestion (contaminated food/water). Exposure to benzene is linked to acute myeloid leukemia and chronic lymphocytic leukemia; breast cancer; and lymphatic and hematopoietic cancer. Benzene is a known human carcinogen.

Bisphenol A (BPA), a building block of polycarbonate plastic, is one of the most widely produced chemicals in the world. It is used in hard plastics, food cans, drink cans, receipts and dental sealants. BPA is ubiquitous. CDC biomonitoring surveys indicate that more than 90 per cent of Americans have the substance in their bodies. BPA is an endocrine disruptor linked to breast and prostate cancer. The International Agency for Research on Cancer has listed BPA as ‘not classifiable as to its carcinogenicity in humans’.

2,2-Bis(bromomethyl)-1,3-propanediol (Technical Grade) (BBMP) is used as a flame retardant in unsaturated polyester resins, for molded products and in the production of rigid polyurethane foam. It is also used as a chemical intermediate in the production of pentaerythritol ethers and other derivatives used as flame retardants.

Butadiene. The major end-use products containing styrene-butadiene and polybutadiene are tyres. Other products include latex adhesives, seals, hoses, gaskets, various rubber products, nylon carpet backings, paper coatings, paints, pipes, conduits, appliance and electrical equipment components, automotive parts and luggage.

Chlorendic acid is used as a flame retardant in polyurethane foams, resins, plasticizers, coatings, epoxy resins and wool fabrics; in the manufacture of alkyl resins for special paints and inks; in the manufacture of polyester resins with special applications in electrical systems, paneling, engineering plastics and paint; and in the manufacture of corrosion-resistant tanks, piping and scrubbers.

Chlorinated paraffins are used as extreme-pressure-lubricant additives in metalworking fluids; as flame retardants in plastics, rubber and paints; to improve water resistance of paints and fabrics; and as a secondary plasticiser in polyvinyl chloride. Small amounts are also used in caulks, sealants, adhesives, detergents, inks, finished leather and other miscellaneous products, and are allowed as an indirect food additive.

Chloroprene. Elastomer polychloroprene (neoprene), a synthetic rubber used in the production of automotive and mechanical rubber goods, adhesives, caulks, flame-resistant cushioning, construction materials, fabric coatings, fibre binding and footwear.

Chromium compounds in wood preservatives increased dramatically from the late 1970s to the early 2000s; however, this use is expected to decrease because of a voluntary phase-out of all residential uses of wood treated with chromated copper arsenate (pressure-treated wood) that came into effect in 2003. Chromium hexavalent compounds are known to be human carcinogens. They were first listed in the *First Annual Report on Carcinogens* (1980). Elemental chromium does not occur naturally; chromium (IV) compounds are highly corrosive and strong oxidizing agents rarely found in nature.

Coal tars and coal pitches are used to make creosote and naphthalene and as a modifier for epoxy-resin surface coatings. Consumer use of creosote has been banned in Europe due to its carcinogenity since 2003 but it is still allowed commercially in some products.

Cumene is used primarily in the manufacture of phenol and acetone (accounting for 98 per cent of all use) and in the manufacture of acetophenone, α-methylstyrene, diisopropylbenzene and dicumylperoxide, and as a thinner for paints, enamels and lacquers.

2,4-diaminotoluene. The primary use has been as an intermediate in the production of 2,4-toluene diisocyanate, which is used to produce polyurethane.

1,2-dibromoethane is used as a chemical intermediate in the manufacture of resins, gums, waxes, dyes and in the manufacture of vinyl bromide, which is used as a flame retardant.

2,3-dibromo-1-propanol is used in the production of flame retardants, and in the production of Tris(2,3-dibromopropyl) phosphate, a flame retardant used in children's clothing and other products. Tris was banned from use in sleepwear in 1977 by the Consumer Product Safety Commission after studies showed that it caused cancer in experimental animals.

1,4-dichlorobenzene has been used primarily as a space deodorant in products such as room deodorisers and toilet deodorant blocks and as a fumigant for moth control.

1,*2-dichloroethane* is used primarily to produce vinyl chloride and as a solvent for resins, rubber and in paint, varnish and finish removers.

Dichloromethane is used as a solvent in paint strippers and removers, in adhesives, as a propellant in aerosols, as a metal cleaning and finishing solvent, and in urethane foam blowing.

Di(2-ethylhexyl) phthalate (DEHP). About 95 per cent of DEHP produced is used as a plasticiser in polyvinyl chloride (PVC) resins. Products typically contain from 1 to 40 per cent DEHP. Plasticised PVC has been used in many consumer items and building products, such as shower curtains, furniture, floor tiles, swimming-pool liners, sheathing for wire and cable and weather stripping. It has been used as a plasticiser in non-PVC materials, including polyvinyl butyral, natural and synthetic rubber, chlorinated rubber, ethyl cellulose and nitrocellulose.

Diglycidyl resorcinol ether is used as a liquid epoxy resin and as a reactive diluent in the production of other epoxy resins used in electrical, tooling, adhesive, casting and laminating applications.

Dimethyl sulfate is used as a methylating or sulfating agent in polyurethane-based adhesives. It was formerly used as a chemical weapon.

1,4-dioxane is used in the manufacture of adhesives, magnetic tape, plastic and rubber.

Dioxins are a group of chemicals formed as unintentional byproducts of industrial processes involving chlorine, such as waste incineration, chemical manufacturing and pulp and paper bleaching. Dioxins include polychlorinated dibenzo dioxins (PCDDs), polychlorinated dibenzofurans (PCDFs) and the polychlorinated biphenyls (PCBs). Exposure is through the ingestion of contaminated foods and, to a lesser extent, dermal contact. Dioxins accumulate in fat cells and degrade very slowly in the environment. The cancer classification depends on the dioxin: 2,3,7,8-TCDD (Agent Orange) is a known human carcinogen; some other dioxins are probable or possible human carcinogens.

Epichlorohydrin is used in the production of numerous synthetic materials, including epoxy, phenoxy and polyamide resins, and polyether rubber used in adhesives. It is also used as a solvent for cellulose, esters, paints and lacquers, and to cure propylene-based rubbers.

Ethylene oxide is mainly used in diethylene glycol, triethylene glycol, polyethylene glycol and urethane polyols.

Formaldehyde. The predominant use (55 per cent of total consumption) is in the production of industrial resins (mainly urea-formaldehyde, phenol-formaldehyde, polyacetal and melamine-formaldehyde resins, including adhesives and binders for composite wood products, pulp and paper products and plastics). Its listing status was changed to known to be a human carcinogen in the *Twelfth Report on Carcinogens* (2011). Formaldehyde can be found in a variety of building and home decoration products (as urea-formaldehyde resins and phenol-formaldehyde resin). It is also used as a preservative and disinfectant. Exposure is through inhalation and dermal contact. Automobile exhaust is the greatest contributor to formaldehyde concentrations in ambient air. Construction materials, furnishings and cigarettes account for most formaldehyde in indoor air. Formaldehyde has caused nasal cancer in rats after long-term exposure; it is linked to leukemia and nasopharygeal cancer in humans. It is a known human carcinogen.

Furan is used in the formation of lacquers and as a solvent for resins.

Inhalable glass wool fibres can generally be classified into two categories based on usage: (1) low-cost, general-purpose fibres typically used for insulation applications, and (2) premium special-purpose fibres used in limited specialized applications. The primary use of glass wool is for thermal and sound insulation. The largest use of glass wool is for home and building insulation in the form of loose wool, batts (insulation in the form of a blanket, rather than a loose filling), blankets or rolls or rigid boards for acoustic insulation. Glass wool is also used for industrial, equipment and appliance insulation. Certain glass wool fibres (inhalable) are reasonably anticipated to be human carcinogens based on evidence of carcinogenicity from studies in experimental animals of inhalable glass wool fibres as a class (defined below) and evidence from studies of fibre properties that indicates that only certain fibres within this class (specifically, fibres that are biopersistent in the lung or tracheobronchial region) are reasonably anticipated to be human carcinogens. There is considerable variation in the physicochemical and biophysical properties of individual glass wool fibres and so carcinogenic potential must be assessed on a case-by-case basis.

Isoprene is used in the production of styrene-isoprene-styrene block polymers and butyl rubber (isobutene-isoprene copolymer).

4,4-methylenebis (2-chloroaniline) is a curing agent for isocyanate polymers and polyurethane prepolymers in the manufacture of castable urethane rubber products and as a curing agent for epoxy.

4,4-methylenedianiline is used in the closed-system production of diisocyanate and polyisocyanates to produce a variety of polymers and resins, such as polyurethane foam, elastomers and isocyanate resins.

Naphthalene is an intermediate in the production of phthalic anhydride, which in turn is an intermediate in the production of phthalate plasticisers, pharmaceuticals, insect repellents and other materials. Naphthalene is commonly found in indoor air quality tests in Europe, largely attributed to flooring products and adhesives.

Nitrobenzene is used in the manufacture of isocyanates, pesticides, rubber chemicals and is used as a solvent for cellulose ester.

2-nitropropane is used as a solvent; it is used in inks, paints, adhesives, varnishes, polymers, resins and synthetic materials.

Pentachlorophenol is used as a wood preservative to prevent fungal decay and insect damage. It was also used as a biocide and was found in ropes, paints, adhesives, leather, canvas, insulation and brick walls. Since 1984, pentachlorophenol has been regulated in the US as a restricted-use pesticide (restricted to certified applicators) for the treatment of utility poles, cross arms, wooden pilings (e.g., wharf pilings), fence posts and timbers for construction.

Polybrominated biphenyls (PBBs) are no longer used in the US. Previously, they were used as flame-retardant additives in synthetic fibres and molded plastics. Their major applications were in thermoplastics, mainly acrylonitrile-butadiene-styrene used in electronic-equipment housings. (In theory PBBs are also banned in Europe but PBDE flame retardants are listed in some health and safety data sheets of common insulation products.)

Polycyclic aromatic hydrocarbons (PAHs) form as a result of incomplete combustion of organic compounds: combustion from wood and fuel in residential heating, coal burners, automobiles, diesel-fueled engines, refuse fires and grilled meats. They are found in coal tar and coal tar pitch, used for roofing and surface coatings. Exposure to these lipophilic substances results from inhalation of polluted air, wood smoke and tobacco smoke, and ingestion of contaminated food and water. PAHs are reasonably anticipated to be a human carcinogen, according to the National Toxicology Program. IARC lists them as probably or possibly carcinogenic.

Polychlorinated biphenyls (PCBs) were used for applications such as transformers, capacitors and heat transfer and hydraulic fluids, inks, flame retardants, adhesives, carbonless duplicating paper, paints, plasticisers, wire insulators, metal coatings and a wide range of plastic products. (PCBs are banned in Europe but remain persistent in the environment through disposal in land fill, etc. If burned PCBs emit dioxins.)

Polybrominated diphenylethers (PBDEs) are used as flame retardants in furniture, computers, electronics, medical equipment and mattresses. Exposure is through inhalation, ingestion and dermal contact. Two of the common commercial formulations, penta- and octa-BDE, have been voluntarily phased out of US production. Deca-BDE continues to be produced. Highly persistent in the environment, they are endocrine disruptors. PBDEs are linked to liver cancer in laboratory animals, but are not classifiable as to carcinogenicity in people.

Propylene oxide is used primarily as a chemical intermediate in the production of other compounds such as polyurethane polyols, propylene glycols, glycol ethers, di- and tripropylene glycols, Polyurethane polyols are used to make polyurethane foams, and propylene glycols are used primarily to make unsaturated polyester resins for the textile and construction industries.

Silica. Exposure of workers to respirable crystalline silica is associated with elevated rates of lung cancer. The link between human lung cancer and exposure to respirable crystalline silica was strongest in studies of quarry and granite workers and workers involved in ceramic, pottery, refractory brick and diatomaceous earth industries. Silicosis, a marker for exposure to silica dust, is associated with elevated lung cancer rates, with relative risks of 2.0 to 4.0.

Styrene is used in the synthesis and manufacture of polystyrene and hundreds of different copolymer. Roughly 99 per cent of the industrial resins produced from styrene can be grouped into six major categories: polystyrene, styrene-butadiene rubber, unsaturated polyester resins (glass reinforced), styrene-butadiene latexes, acrylonitrile-butadiene-styrene and styrene-acrylonitrile. Polystyrene is used extensively in the manufacture of plastic packaging, thermal insulation in building construction and refrigeration equipment, and disposable cups and containers. Styrene polymers and copolymers are also used to produce various corrosion-resistant tanks and pipes, various construction items, carpet backings, house paints, computer printer cartridges, insulation products, wood-floor waxes and polishes, adhesives and putties. Styrene-butadiene rubber is the most widely used synthetic rubber in the world. Over 70 per cent of styrene-butadiene rubber is consumed in the manufacture of tyres and tyre products; however, non-tyre uses are growing, with applications including conveyor belts, gaskets, hoses, floor tiles, footwear and adhesives. Styrene exposure to the general population can occur through environmental contamination. For the non-smoking general population, inhalation of indoor air and ingestion of food result in the highest daily styrene intakes. Styrene has been measured in outdoor air, but higher levels generally are found in indoor air, drinking water, groundwater, surface water, soil and food. Styrene can be emitted to the air from off-gassing of building materials and consumer products, and cigarette smoking.

Styrene oxide is used as a reactive diluent for epoxy resins and in cross-linked polyesters and polyurethanes. Styrene oxide has been used as a raw material for the production of 2-phenylethanol (oil of roses) used in perfumes and in the treatment of fibres and textiles.

Toluene diisocyanates is used primarily to manufacture flexible polyurethane foams for use in furniture, bedding and automotive and airline seats. Toluene diisocyanate–based rigid polyurethane foam is used in household refrigerators and for residential sheathing or commercial roofing in board or laminate form; spray-in rigid foam is used as insulation. Polyurethane-modified alkyds contain approximately 6 to 7 per cent isocyanate, mostly toluene diisocyanates, and are used as coating materials, such as floor finishes, wood finishes and paints. Moisture-curing coatings are used as wood and concrete sealants and floor finishes. They are used in adhesive and sealant compounds and in automobile parts, shoe soles, rollerskate wheels, pond liners and blood bags.

Trichloroethylene is used as a modifier for polyvinyl chloride polymerization. Five main industrial groups use trichloroethylene in vapor or cold degreasing operations: furniture and fixtures production, fabricated metal products, electrical and electronic equipment, transport equipment, and miscellaneous manufacturing industries. Trichloroethylene is used as a solvent in the rubber industry, adhesive formulations, dyeing and finishing operations, printing inks, paints, lacquers, varnishes, adhesives and paint strippers.

Tris(2,3-dibromopropyl) phosphate was used in the US for making flame-retardant children's clothing. In 1977, the Consumer Product Safety Commission banned the use of Tris(2,3-dibromopropyl) phosphate in children's clothing and in fabric, yarn and fibre when intended for use in such clothing. Tris(2,3-dibromopropyl) phosphate has also been used as a flame retardant in polyurethane foams for cushioning, insulation, furniture and automobile and aircraft interior parts, as well as in polystyrene foam, acrylic carpets and sheets, water flotation devices, polyvinyl and phenolic resins, paints, lacquers, paper coatings, styrene-butadiene rubber and latexes.

Urethane. The primary use of has been as a chemical intermediate in preparation of amino resins (IARC 1974). The process involves a reaction with formaldehyde to give hydroxymethyl derivatives that are used as cross-linking agents in permanent-press textile treatments designed to impart wash-and-wear properties to fabrics.

4-vinyl-1-cyclohexene diepoxide is used as a reactive diluent for other diepoxides and for certain epoxy resins derived from bisphenol A and epichlorohydrin.

Vinyl halides. Vinyl bromide is used primarily in the production of polymers and copolymers. It is used in polymers as a flame retardant and in the production of monoacrylic fibres for carpet-backing material. Combined with acrylonitrile as a co-monomer, it is used to produce fabrics and fabric blends used in sleepwear (mostly children's) and home furnishings. Vinyl chloride is used almost exclusively by the plastics industry to produce polyvinyl chloride (PVC) and copolymers. PVC is a plastic resin used in many consumer and industrial products, including automotive parts and accessories, furniture, medical supplies, containers, wrapping film, battery cell separators, electrical insulation, water-distribution systems, flooring, windows, videodiscs, irrigation systems and credit cards. More than 95 per cent of all vinyl chloride monomer produced is used to make PVC and its copolymers.

Vinyl chloride previously was used as a refrigerant, as an extraction solvent and in aerosol propellants, but these uses were banned in 1974 because of vinyl chloride's carcinogenic effects. Exposure is largely occupational and results from inhalation, ingestion or dermal contact. Exposure is very low in the general population. Exposure to vinyl chloride is linked to the development of liver cancer and weakly associated with brain cancer. Vinyl chloride is a known human carcinogen. Vinyl fluoride is used primarily in the production of polyvinyl fluoride and other fluoropolymers. Polymers of vinyl fluoride are resistant to weather and exhibit great strength, chemical inertness and low permeability to air and water. Polyvinyl fluoride is laminated with aluminum, galvanized steel and cellulose materials and is used as a protective surface for the exteriors of residential and commercial buildings. Polyvinyl fluoride laminated with various plastics has been used to cover walls, pipes and electrical equipment and inside aircraft cabins.

Wood dust is produced in woodworking industries as a byproduct of the manufacture of wood products; it is not usually produced for specific uses. One commercial use for wood dust is in wood composts. 'Industrial roundwood' refers to categories of wood not used for fuel, which include sawn wood, pulpwood, poles and pit props and wood used for other purposes, such as particleboard and fibreboard. Many case reports and epidemiological studies have found a strong association between exposure to wood dust and cancer of the nasal cavity and paranasal sinuses.

Appendix B

Useful organisations in UK, Europe, USA and internationally

UK associations/NGOs

UK Indoor Environment Group (UKIEG) Has approximately 250 registered members including medics, toxicologists, architects, designers, appliance manufacturers, academics, regulators, researchers, chemists, engineers, building managers, environmental health professionals, social scientists and others working in fields connected with the built environment.
www.ukieg.org/

Alliance for Sustainable Building Products (ASBP)
www.asbp.org.uk

The Bhopal Medical Appeal
http://bhopal.org/

British Lung Foundation
www.blf.org.uk/support-for-you/air-pollution

Cancer Research UK
www.cancerresearchuk.org/about-cancer/causes-of-cancer/air-pollution-radon-and-cancer/how-air-pollution-can-cause-cancer

Centre for Alternative Technology
www.cat.org.uk

COMEAP Secretariat Public Health England Centre for Radiation, Chemical and Environmental Hazards, Chilton, Oxon, OX11 0RQ

Committee on the Medical Effects of Air Pollutants (COMEAP)
www.gov.uk/government/groups/committee-on-the-medical-effects-of-air-pollutants-comeap

Energy Institute Health Technical Committee
Energy Institute
61 New Cavendish Street, London, W1G 7AR, UK
t: +44 (0) 20 7467 7100;
f: +44 (0) 20 7255 1472

Indoor Air Quality Association (IAQ UK)
www.iaquk.org.uk/

Publications on air quality and emissions mainly concerned with occupational health of oil and gas industry employees

Clean Air in London
http://cleanair.london/contact/
http://cleanair.london/hot-topics/indoor-air-quality-can-be-worse-than-outdoor/

Pesticides Action Network UK
www.pan-uk.org/

Womens Environmental Network
www.wen.org.uk/#home
Published guide to healthy and environmental flooring

UK Business Council for Sustainable Development (UKBCSD)
www.ukbcsd.org.uk/

UK commercial organisations

Allergy Cosmos Supplies air filtration equipment
www.allergycosmos.co.uk/commercial-air-filtration/concern/voc-pollution

Allergy UK
www.allergyuk.org/
Helpline 01322 619 898

Building Research Establishment (BRE) BREEAM and Home Quality Mark and air quality testing
www.bre.co.uk/page.jsp?id=720

Enviroact Consultants
http://ibe.sagepub.com/

Gale and Snowden Architects Building Biology
www.ecodesign.co.uk/approach/healthy-buildings/

Healthy Buildings International
http://hbi.co.uk

Safe Air Quality.com
www.safeairquality.com/about-us/

Waverton Analytics Home Air Check™
http://homeaircheck.co.uk/customer-care
Commercial and domestic indoor air quality monitoring

UK university research centres

Derby University Masters Course in Sustainable Architecture and Healthy Buildings
www.derby.ac.uk/courses/postgraduate/sustainable-architecture-healthy-buildings-msc/

Institute of Environment and Health Cranfield University
www.cranfield.ac.uk/about/media-centre/news-archive/news-2015/improving-air-quality-in-buildings.html

Mackintosh Environmental Architecture Research Unit (MEARU)
www.gsa.ac.uk/research/research-centres/mearu/

University College London
www.bartlett.ucl.ac.uk/iede/research/themes/healthy-buildings
www.activebuildings.co.uk/news

University of Leeds
www.engineering.leeds.ac.uk/people/civil/staff/c.j.noakes
www.refresh-project.org.uk/about/

European Union

Institut für Baubiologie GmbH (IBR)
www.baubiologie-ibr.de

Natureplus
www.natureplus.org

Sinphonie
www.sinphonie.eu/

Sentinel Haus
www.sentinel-haus.eu/en/

The Institute of Building Biology + Sustainability (IBN)
www.baubiologie.de/international/institute

Umweltbundesamt
www.umweltbundesamt.de/en/topics/health/environmental-impact-on-people/indoor-air-hygiene
www.umweltbundesamt.de/en/topics/health/commissions- *AGBB/CERTECH* working-groups/indoor-air-hygiene-commission
www.umweltbundesamt.de/en/topics/health/commissions-working-groups/committee-for-health-related-evaluation-of-building

USA

American Cancer Society
www.cancer.org/
www.cancer.org/cancer/cancercauses/othercarcinogens/intheworkplace/formaldehyde
www.cancer.org/cancer/cancercauses/othercarcinogens/generalinformationabout carcinogens/known-and-probable-human-carcinogens

American Industrial Hygiene Association Membership association, publishes research in IAQ and VOCs
www.aiha.org/Pages/default.aspx

Ashrae
www.ashrae.org/resources—publications/bookstore/indoor-air-quality-guide

Building Ecology.com
www.buildingecology.com/

Building Green
www2.buildinggreen.com/

Green Building.com
www.greenbuilding.com/knowledge-base/indoor-air-quality

Healthy Building Network
www.healthybuilding.net/

Healthy Building Science Consultants
http://healthybuildingscience.com/environmental-consultants/#

Healthy House Institute
www.healthyhouseinstitute.com/a-799-Choosing-Healthy-Building-Materials

Healthy Materials Lab The New School Parsons School of Design, New York
www.healthymaterialslab.org/bill-walsh/

National Academies of Science, Engineering and Medicine
http://iom.nationalacademies.org/Activities/PublicHealth/Health-Risks-Indoor-Exposure-ParticulateMatter/2016-FEB-10.aspx

The Pharos Project
www.pharosproject.net/

US Environmental Protection Agency (US EPA)
www.epa.gov/

US Green Building Council/LEED (USGBC)
www.usgbc.org/credits/eq22

International

International Society of Indoor Air Quality and Climate (ISIAQ)
www.isiaq.org/

Air Infiltration and Ventilation Centre
www.aivc.org/

Health Care Without Harm
https://noharm-europe.org/

Health and Environment Alliance (HEAL)
www.env-health.org/about-us/

International Centre for Indoor Environment and Energy (ICIEE) Denmark
www.iciee.byg.dtu.dk/About-ICIEE/Mission—Vision

International Society of Indoor Air Quality and Climate (ISIAQ)
www.isiaq.org/

World Health Organization
www.who.int/hia/housing/en/

Index